Real World Application Of Industrial Engineering Vol.-1

SUNIL RANA

ISBN: **1495428109**
ISBN-13: **978-1495428104**

DEDICATION

This book is dedicated to my parents, my wife and my daughter for motivating me and supporting me in the whole process of writing this book. Also this book is dedication to my **Alma-Mater Motilal Nehru Regional Engineering College, Allahabad (U.P), India now M.N.N.I.T.** and to all Industrial Engineering graduates & professionals all over the globe and future Industrial Engineers.

CONTENTS

ACKNOWLEDGMENTS

I am thankful to my professors in college, my mentors and colleagues in various companies I had worked. These are the people who have helped me grow as an I.E. professional and have taught me the real world applications of Industrial Engineering.

PREFACE

This book is targeted at professionals and students who have background knowledge of Industrial Engineering concepts or are pursuing studies in Industrial Engineering. Therefore it is expected that those who are referring to the real world problems and solutions provided in this book are well versed in basic mathematics, science and I.E. concepts.

The examples provided in this book are the ones on which I have worked individually or in teams to arrive at a solution in various companies I have worked so far. The solutions provided may not be the best but are the ones that were workable within applicable cost, quality and time constraints. Therefore don't limit your thought process and if need be try to view the same problem from your own perspective to arrive at a different alternative for the problem at hand.

This book will also highlight the expectations of the corporate world from the Industrial Engineers, the software systems being used by I.E., additional qualification, certification I.E.'s can pursue to upgrade their knowledge and skills to make them saleable in the competitive job environment.

Apologies for the brevity and grammatical errors you may encounter going through the book.

1 APPLICATION OF I.E. IN MANUFACTURING SECTOR

The role of Industrial Engineer in manufacturing industry is an amalgamation of Human Resource , Finance, Inventory control and Change agents. An I.E. is supposed to do Manpower Planning, Budgeting of key value drivers & implementing industry best practices.
The responsibility of I.E. is to solve problems or opportunities of improvement, in following areas:

- Job Evaluation and Wage system
- Work time, Motion and Method Study
- Material management
- Ergonomics
- Projects and developing Systems
- Value Engineering
- WCM
- Costing
- Logistic
- Finance
- Accounting
- Lean
- SCM
- HR and Administration
- Six Sigma

There are following categories of jobs that industrial engineers are engaged in manufacturing:

1) Basic Functions:

The tasks which are done on day to day basis e.g. 1) Data analysis to categorize items and Maintain the record of inventory items issued especially category "A" items i.e. items with high value, low quantity and low turnaround, 2) Time study of tasks in order to establish standards to calculate manpower requirement in different areas and standard rates in case of outsourced jobs. 3) Life Cycle Cost of the equipment

Industrial Engineering concepts to be utilized in such cases are:

ABC Analysis:

Where high value items are categorized as "A" items, they account for only 10% of the total quantity but 70-80% of the total inventory carrying cost. Such items needs high control and good amount of documentation and justification from the department asking for it.

In inventory control systems like Maximo such items are designated as "Critical" items and normally while issuance, the requisition has to pass through 2-3 stages of high level approval.

Time and Motion study, ILO / Company policies defined allowances :

The jobs under study are normally long cycle and hence doesn't require use of stop watches as such. The task involve transition from one activity to another which has to be jotted down and Man-Heads involved in the activity needs to be noted. Given below is the sample format (activities have been omitted intentionally) of one such study:

S1-Sample time study:

Sno.	Description	Start	End	Duration {= (Start Time - End Time) x 24 x 60}	Man-Heads	Man-minutes
1	XXXXX	10:00	10:13	13.00	2	26
2	YYYYY	10:13	11:00	47.00	3	141
3	ZZZZZ	11:00	11:20	20.00	1	20
4	VVVV	11:20	12:00	40.00	2	80
	Total Normal Man-minutes					267
	Additional time for Basic Allowance (9%)					24.03
	Additional time for Heat allowance (8%)					21.36
	Total Standard Man-minutes					312.39

S2- Table of Labor Allowance (As recommended by ILO):

ILO Recommended Allowances for PFD				
A.	**Constant allowances:**			%
	1	Personal allowance		5
	2	Basic fatigue allowance		4
B.	**Variable allowances:**			
	1	Standing allowance		2
	2	Abnormal position allowance:		
		a.	Slightly awkward	0
		b.	Awkward (bending)	2
		c.	Very awkward (lying, stretching)	7
	3	**Use of force, or muscular energy (lifting, pulling, or pushing):**		
		Weight lifted, pounds:		
			5	0
			10	1
			15	2
			20	3
			25	4
			30	5
			35	7
			40	9
			45	11
			50	13
			60	17
			70	22

4	**Inadequate light:**		
	a.	Slightly below recommended	0
	b.	Well below	2
	c.	Quite inadequate	5
5	**Atmospheric conditions (heat and humidity) - variable**		0-100
6	**Close attention:**		
	a.	Fairly fine work	0
	b.	Fine or exacting	2
	c.	Very fine or very exacting	5
7	**Noise level:**		
	a.	Continuous	0
	b.	Intermittent - loud	2
	c.	Intermittent - very loud	5
	d.	High-pitched - loud	5
8	**Mental strain:**		
	a.	Fairly complex process	1
	b.	Complex or wide span of attention	4
	c.	Very complex	8
9	**Monotony:**		
	a.	Low	0
	b.	Medium	1
	c.	High	4
10	**Tediousness:**		
	a.	Rather tedious	0
	b.	Tedious	2
	c.	Very tedious	5

Life cycle costing of the equipment:

The Goal of LCC is to choose the best option of the alternatives available all of which are meeting technical specification and requirement of the company. The goal is to spend the amount of capital that will produce the lowest possible life cycle cost. The costing analysis is divided into following steps:

1 Identify the feasible alternatives

2 Analyze the total lifetime of events for physical assets. Include future applicable activities associated with R & D, production, construction, Installation, Commissioning, operation, Maintenance, and Disposal etc. Prepare Cost Break up structure (CBS).

3 Setup a model to define the cost factors and estimate the relationship (Labor rates, Mandated profit Margins, Fuel consumption Rates)

4 Work up the cost of each of the life cycle elements

5 Account for inflation and learning curve to set the required accuracy.

6 Discount all the Estimated cost to a base period (using concept of sinking fund, present value (PV) and capital recovery)

7 Identify high cost contributors (search for underlying cause and eliminate or mitigate)

8 Calculate the final LCC by using appropriate cost model (consider the characteristic of asset and management approach used for LCC analysis)

9 Perform the risk analysis.

10 Recommend a solution.

Below is an example given for Life cycle costing:

Factor Calculation:

i = Discount Rate =	12.00%
f = Inflation Rate =	5.00%

Year (t)	Present Worth Factor (B) $(l+0.01i)^{-t}$	Uniform Present Worth Factor (C = Cumulative)	Inflation Factor (D) $(l+0.01f)^{t}$	Present Worth Factor Coosidering Inflation = B x D $(l+0.01i)^{-t} * (l+0.01f)^{t}$	Amortisation Factor 1/C	Uniform Present Worth Factor with Inlation (C)
1	0.8929	0.8929	1.0500	0.9375	1.1200	0.9375
2	0.7972	1.6901	1.1025	0.8789	0.5917	1.8164
3	0.7118	2.4018	1.1576	0.8240	0.4163	2.6404
4	0.6355	3.0373	1.2155	0.7725	0.3292	3.4129
5	0.5674	3.6048	1.2763	0.7242	0.2774	4.1371
6	0.5066	4.1114	1.3401	0.6789	0.2432	4.8160
7	0.4523	4.5638	1.4071	0.6365	0.2191	5.4525
8	0.4039	4.9676	1.4775	0.5967	0.2013	6.0492
9	0.3606	5.3282	1.5513	0.5594	0.1877	6.6086
10	0.3220	5.6502	1.6289	0.5245	0.1770	7.1331
11	0.2875	5.9377	1.7103	0.4917	0.1684	7.6248
12	0.2567	6.1944	1.7959	0.4610	0.1614	8.0857
13	0.2292	6.4235	1.8856	0.4321	0.1557	8.5179
14	0.2046	6.6282	1.9799	0.4051	0.1509	8.9230
15	0.1827	6.8109	2.0789	0.3798	0.1468	9.3028
16	0.1631	6.9740	2.1829	0.3561	0.1434	9.6589
17	0.1456	7.1196	2.2920	0.3338	0.1405	9.9927
18	0.1300	7.2497	2.4066	0.3130	0.1379	10.3057
19	0.1161	7.3658	2.5270	0.2934	0.1358	10.5991
20	0.1037	7.4694	2.6533	0.2751	0.1339	10.8741

Life Cycle Costing:

LCC CALCULATION SHEET - PRESENT WORTH FACTOR METHOD

Column C	Column D	Column E	Column F	Column G
DESCRIPTION		**ALTERNATE 1**		**ALTERNATE 2**
NAME OF EQUIPMENT / FACILITY :		**XXXX**		**YYYY**
NO REQUIRED TO MEET THE PRODUCTION REQUIREMENTS		**6 Nos**		**4 Nos**
EXPECTED LIFE	20	Years	20	Years
ai = Discount Rate =	12%	Percent	12%	Percent
f = Inflation Rate =	5%	Percent	5%	Percent

1.0	**INITIAL COST**	**ESTIMATED COST**	**PRESENT WORTH**	**ESTIMATED COST**	**PRESENT WORTH**
a	Purchase cost (For 6 Nos)	1,96,00,000	1,96,00,000	1,85,00,000	1,85,00,000
b	Erection cost & Commi.cost (15 % of Purchase Cost)	29,40,000	29,40,000	27,75,000	27,75,000
c	Misc.expenses (5 % of Purchase Cost)	9,80,000	9,80,000	9,25,000	9,25,000
	Total =	**2,35,20,000**	**2,35,20,000**	**2,22,00,000**	**2,22,00,000**
2.0	**Single Expenditure &Spare Parts Consumed Intermittently for both the Alternates (Enter wherever necessary Excluding Spare Parts which are consumed every Year)**				
a	Single Expenditure / Spare Parts required in Year 1	75,000	70,313	50,000	46,875
b	Single Expenditure / Spare Parts required in Year 2	0	0	0	0
c	Single Expenditure / Spare Parts required in Year 3	75,000	61,798	50,000	41,199
d	Single Expenditure / Spare Parts required in Year 4	8,52,000	6,58,150	5,68,000	4,38,766
e	Single Expenditure / Spare Parts required in Year 5	75,000	54,315	50,000	36,210
f	Single Expenditure / Spare Parts required in Year 6	0	0	0	0
g	Single Expenditure / Spare Parts required in Year 7	75,000	47,738	50,000	31,825
h	Single Expenditure / Spare Parts required in Year 8	25,32,000	15,10,894	16,88,000	10,07,262
i	Single Expenditure / Spare Parts required in Year 9	75,000	41,957	50,000	27,971
j	Single Expenditure / Spare Parts required in Year 10	0	0	0	0
k	Single Expenditure / Spare Parts required in Year 11	75,000	36,876	50,000	24,584
l	Single Expenditure / Spare Parts required in Year 12	8,52,000	3,92,731	5,68,000	2,61,821
m	Single Expenditure / Spare Parts required in Year 13	75,000	32,411	50,000	21,607
n	Single Expenditure / Spare Parts required in Year 14	0	0	0	0
o	Single Expenditure / Spare Parts required in Year 15	75,000	28,486	50,000	18,991
p	Single Expenditure / Spare Parts required in Year 16	25,32,000	9,01,580	16,88,000	6,01,053
q	Single Expenditure / Spare Parts required in Year 17	75,000	25,036	50,000	16,691
r	Single Expenditure / Spare Parts required in Year 18	0	0	0	0
s	Single Expenditure / Spare Parts required in Year 19	75,000	22,005	50,000	14,670
t	Single Expenditure / Spare Parts required in Year 20	0	0	0	0
	Total =	**75,18,000**	**38,84,287**	**50,12,000**	**25,89,525**

Sunil Rana: Present Worth is calculated as "Estimated Cost x Present worth Factor with inflation for that year = 75000 x 0.8789"

Sunil Rana: Estimated cost is something which is known

		ESTIMATED COST	PRESENT WORTH	ESTIMATED COST	PRESENT WORTH
3.0	**Salvage Value**	19,60,000	2,03,187	18,50,000	1,91,784
4.0	**Actual Operation & Maintenance costs every year:**				
a	*Consumable spares every year*	2,70,000	74,266	1,80,000	49,511
b	*Power cost*	1,57,24,800	43,25,244	2,09,66,400	57,66,993
c	*Air cost*	60,000	16,504	28,000	7,702
d	*Water cost*	1,20,000	33,007	80,000	22,005
e	*Lubricants cost*	60,000	16,504	40,000	11,002
f	*Man power cost*	5,00,000	1,37,529	3,00,000	82,518
g	*Misc.expenses*	0	0	0	0
	Total =	1,67,34,800	46,03,054	2,15,94,400	59,39,730
5.0	**Life cycle cost (1 +2 +4)**		**3,20,07,341**		**3,07,29,255**
6.0	**Less Salvage Value (3) =**		**3,18,04,154**		**3,05,37,471**
	Total Life Cycle Cost :		**3,18,04,154**		**3,05,37,471**

Sunil Rana:
Present worth of salvage Value is calculated as:
Purchase Cost x 10% x Present worth factor @ 20 yrs = 19600000 x 10% x .1037 = 203187

Sunil Rana: Present Worth of Ops & Maint. Cost is calculated as: Estimated cost x Present worth considering inflation @ 20 yrs = 270000 x .2751 = 74266

2) Annual Budgeting:

Industrial Engineering dept. acts as a bridge between Operations and Finance functions. Finance dept. depends on I.E. for sanctity check of the requirements raised by Operations.

The concept of Zero Based Budgeting is deployed to plan and budget the resources for the financial year.

Definition of Zero Base Budgeting

A method of budgeting whereby all activities are reevaluated each time a budget is set. Discrete levels of each activity are valued and a combination chosen to match funds available. –CIMA Terminology .

Comparison between ZBB and Traditional Budgeting:

Zero Based Budgeting	Traditional Budgeting
No reference to previous year expenditure	Previous year actuals and estimates are referred
Budget request has to be justified in complete detail by each section manager starting from the Zero-base	Factors like inflation etc. are adjusted to previous estimates to arrive at the figures of current year's budget
The Zero-base is indifferent to whether the total budget is increasing or decreasing	Managers, in traditional method were able to manipulate their budget estimates. Top management decides on the

	allocation of funds
Zero Based Budgeting	
Advantages	**Disadvantages**
Efficient Resource Allocation based on needs and benefits	**Difficult to define decision units and decision packages**
Managers forced to find cost effective ways to improve operation	**The R&D department is threatened whereas the production department benefits**
Increases staff motivation on account of greater initiative and responsibility in decision-making.	**Necessary to train managers**

The role of I.E. is to consolidate requirements from the respective allocated department, organize, coordinate budget meetings with the concerned departments. And finally take agreement from all Business Packets within Business unit / Department on the agreed budget and hand it over to the Finance and Statistics department.

3) Improvement Initiatives: Improvement initiatives range from small improvement projects (Point Kaizen or System kaizen) to Turn Key projects (Major Reengineering, Six Sigma Projects).

Given Below are examples of nature of projects that can be assigned to I.E. team by Top Management.

1) Product recovery from a process. (System Kaizen)
2) Scrap reduction and recycling. (System Kaizen)
3) Reduction in maintenance cost and Mean Time To Repair (Six Sigma)

The aim of all these projects is to achieve WCM (World Class Manufacturing) philosophy of the company in order to survive and perform well in the competitive environment.

2 WORLD CLASS MANUFACTURING

World Class Manufacturing (WCM) practices are meant to promote excellence in manufacturing as a means of delighting our customers, employees and other stakeholders on sustainable basis. The aim of the system is:

- Zero Defects
- Zero Breakdowns
- Zero Accidents
- Zero Customer Complaints
- Increase in Plant Efficiency
- Reduction in Manufacturing Costs
- Delivery of Quality Product on Time to the Customer
- Boosting Morale of Employees

World Class Manufacturing : 8 dimensions

1. Waste (Muda) Elimination
2. Work Environment (5-S)
3. JIT / Supply Chain Management
4. Equipment Effectiveness / TPM
5. Customer Driven : Internal & External
6. Quality First : SQM and Best Practices
7. Liaison Team Force and Skill Development
8. Information Systems / BPR, Technology and Cash Flows

Equipment Effectiveness (Total Productive Maintenance) : 8 Pillar

1. Focused Improvement
2. Autonomous Maintenance

3. Planned Maintenance
4. Quality Maintenance
5. Development Management
6. Education & Training
7. Office TPM
8. Safety & Environment

Key Manufacturing Focused Areas: 14 Sub –Committees

KMFAs:

1. Work Environment & Waste Elimination
2. Autonomous (Self) Maintenance
3. Focused Improvement and innovation
4. Visual Management & Control
5. Planned Maintenance
6. Strategic Quality Management and Bench Marking
7. Initial Flow Control / Upstream Management
8. Market Orientation & Customer Driven
9. JIT / Supply Chain Management
10. Safety, Hygiene and Pollution control
11. Liaison, Team Force and Skills Development
12. Information and Systems
13. Technology
14. Cash Flow and Cost Drivers

It's important from I.E. point of view to have awareness about all the new

philosophies in order to implement in the concerned / dedicated areas.
Given below is the elaboration on few of these strategies with examples for better understanding.

1st Dimension of WCM : Waste (Muda in Japanese) Elimination:

Easiest way of remembering 7 types of wastes:

Mr. Tim Wood

Transportation

Inventory

Motion

Waiting

Overproduction

Over processing

Defects

- **Transportation**
  - Movement of men, material , semi-finished component from one station to another adds no value to the product

Transportation waste can be eliminated by shortening the distance between work stations, placing material in dedicated area near the work station.

In order to eliminate waste in travel time, the best way is to make spaghetti diagram of the work flow in the area. Then find out the lines which goes back to the previous station, these are the movements which are occurring most likely due to absence of a similar finishing station at the end of the product flow. Try to get a new work station at the end of work flow if there is no cost constraint or else try to minimize the travel time by shifting the station to a mid-location.

Given below is an example of spaghetti diagram of a product in fabrication shop:

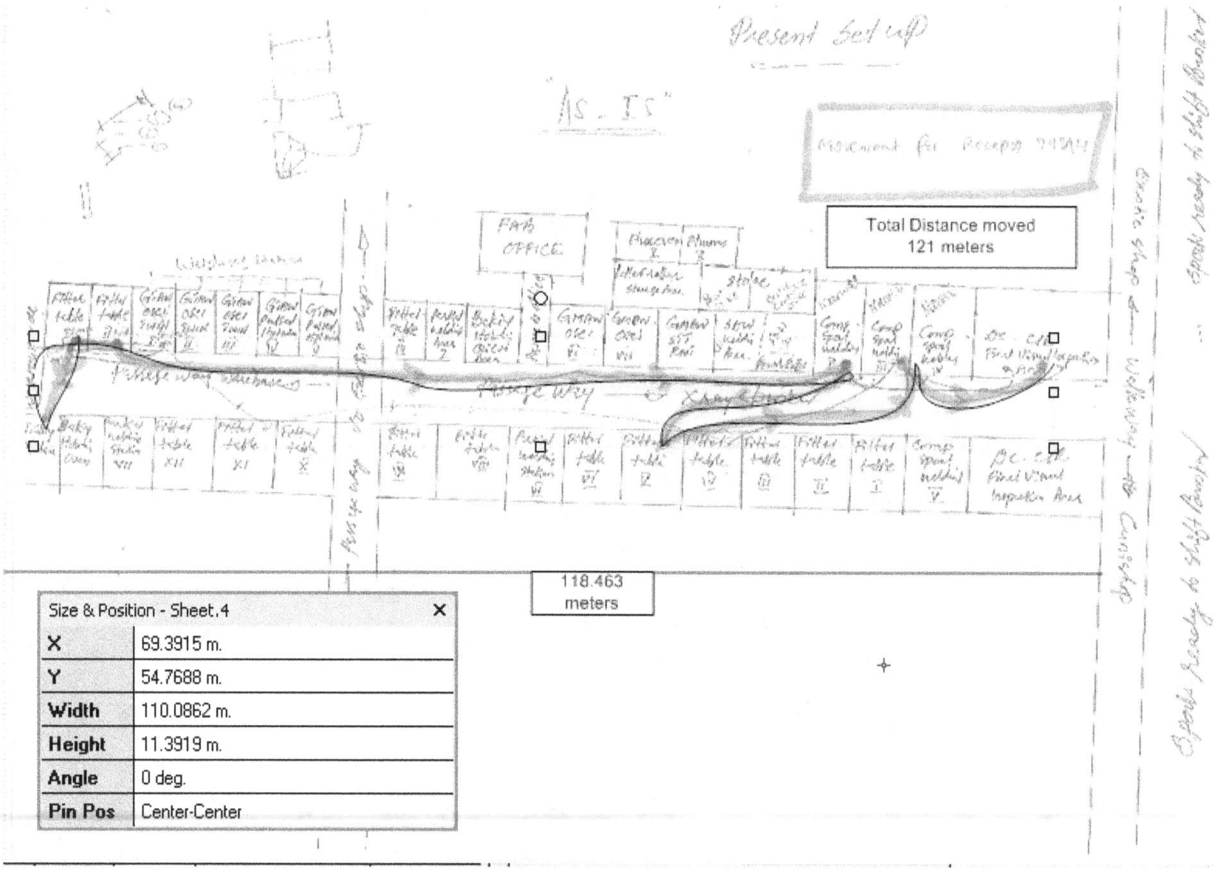

The backward movement of product can be avoided by shifting the station ahead in the flow.

- **Inventory**
    - Excess raw material / semi-finished components lying before start of the operation consumes productive floor or warehouse space, delays the identification of problem areas, and inhibits communication

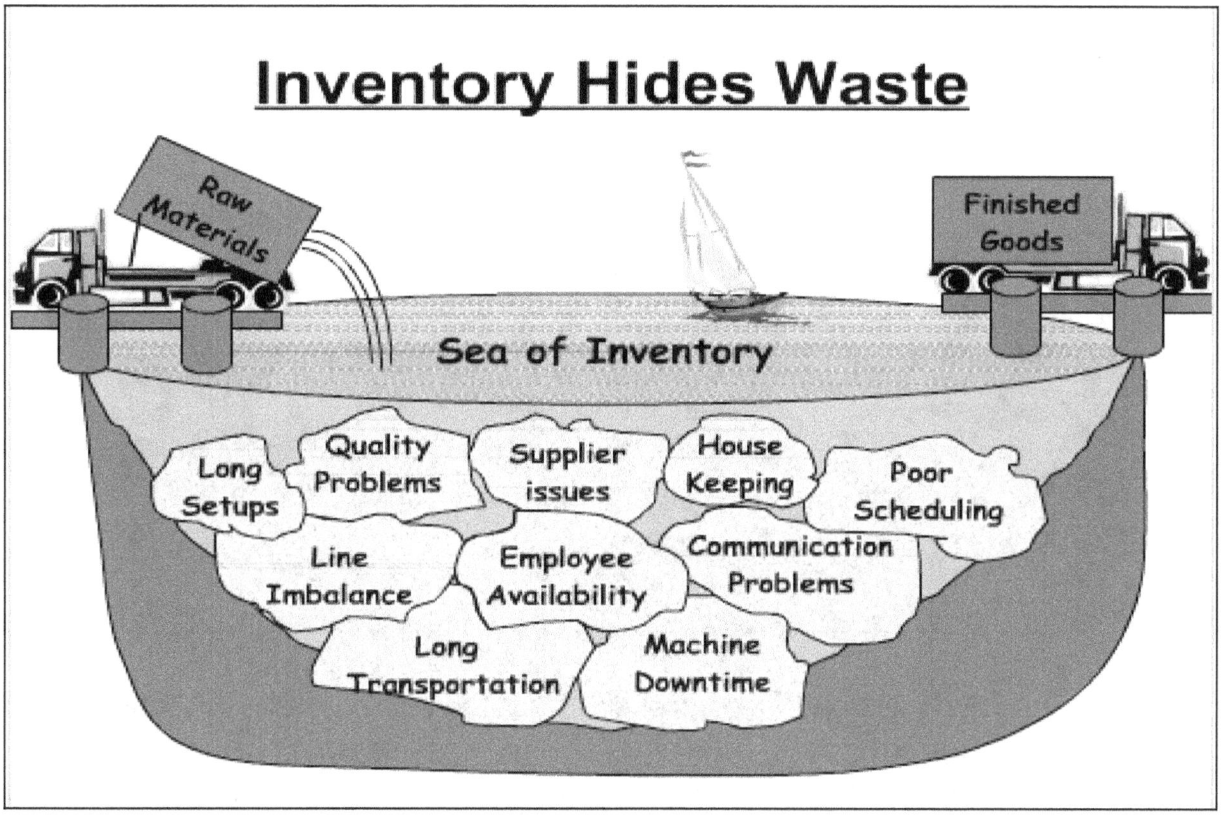

In manufacturing, the spares & consumables can cost the company fortune if it is not managed properly. A company may keep on buying same spares and consumables again and again because of lack of control & lost material. The good thing about inventory is that it hides all the problems by bringing in a substitute but these benefits can help only in the short term. In long term all the spending on excess inventory can erode company's bottom line and profitability because 1) There's a depreciation cost involved 2) if left unused they'll only sell as 3) The valuable space that could have been used for improving production goes waste occupied by heaps of inventory.

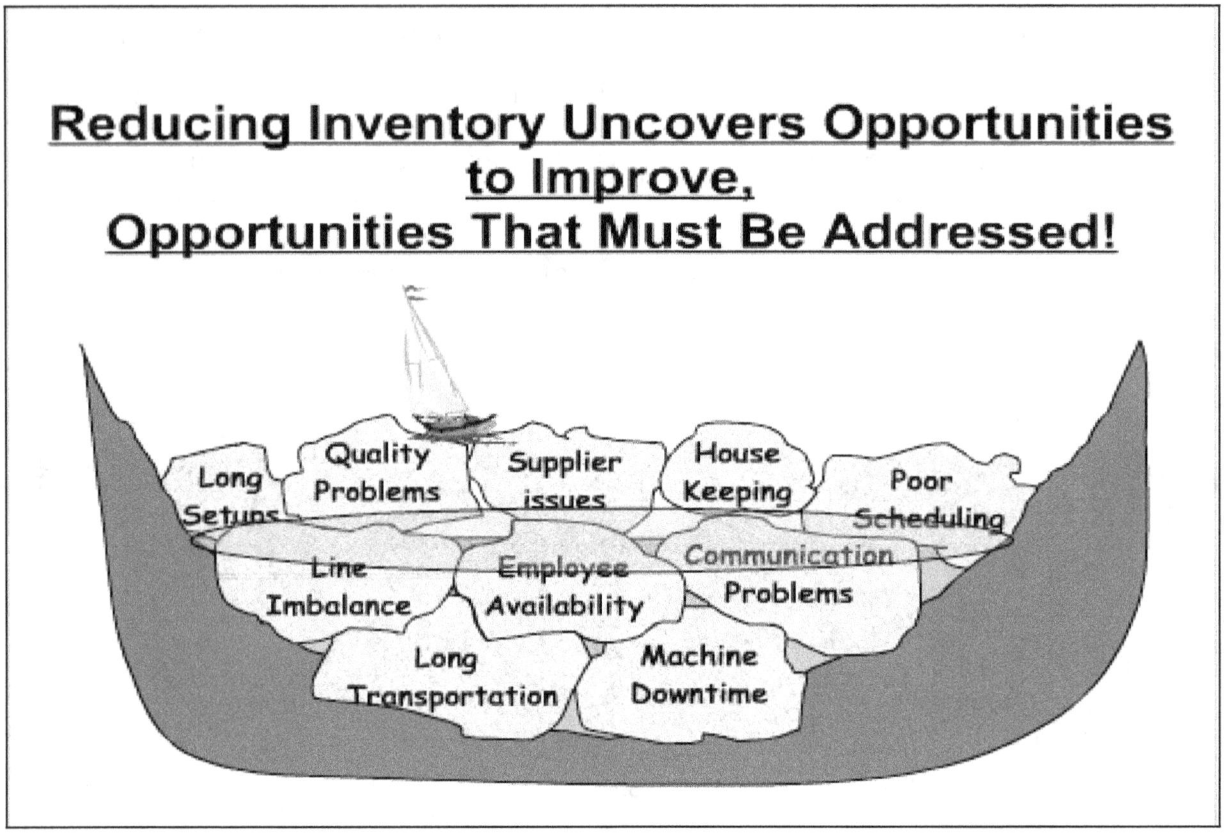

Reducing inventory will lead to uncovering of hidden problems and hence the opportunity for improvement. But all this needs top management support. This cannot be achieved in silos, the CEO or Chief manufacturing officer (CMO) of the company should be the champion and oversee all the efforts across all dept.'s. Just by opening another dept. to work towards improvement will lead the organization nowhere. Sooner or later the program for improvement will come to an end.

Now the bigger question is, can the inventory be reduced to zero by applying Just in time concepts? The answer is no. There has to be some buffer that a company needs to carry in order to avoid lost production, lost sales and lost profits due to unavailability of the spares at the opportune time.

Theory of constraints is what works in the real world. There are times when you have political, social and environmental factors working to shatter the Just-in time concepts. Examples: The Truck carrying supply of spares is stuck at the state border due to lack of some paper work, roads blocked due to heavy rain or possible strike by some organization, the delivery didn't reach on time due to huge traffic due to religious congregation.

All these factors are out of company control and this needs to be built in while planning for the inventory in annual budgeting.

- **Motion**

  Waste of Man-Hours for getting tools from store/far off location, Bending, stretching, walking, lifting due to poor workplace layout and working methods.

The waste in motion is not evident on the job at an instant because it seems so natural and perfect in the given circumstances to execute the task in the sequence it is being done. It has implications in the medium term which are visible in the productivity figures of the department.

Working in awkward position, lifting heavy loads leads to fatigue and musculoskeletal injuries in the long term. Increasing medical expenses and insurance cost by the company on workforce. The effort which otherwise should have gone in productive job goes waste in non-value added task.

Every effort shall be made to make the work easier for the workforce to reap the benefits in the long term.

- **Waiting**
    - For tools, instructions, drawings , inspection etc.

The waiting waste happens because of poor planning, poor coordination between different departments. There can't be nothing worse for a supervisor to see his team sitting idle because the material didn't reach on time, drawing of the component is missing, machine has broken down. Ultimately at the end of the shift a supervisor is responsible to give the planned output.

- **Overproduction**
    - Producing components long before required for next operation.

This generally happens when one of the station or process becomes a bottleneck in the whole supply chain. It becomes difficult for it to process what it has got from the previous station / process. This requires quick action otherwise the production can stall and can have repercussions which would last days to months.

Examples:

1) Finished product is ready but there are not much sales order leading to stacking of the products at distributor / dealers warehouse.
2) IT Servers down leading to holding of information from reaching on time.

- **Over processing**
    - Doing more than necessary to produce an effectively functioning product, extra set up steps, over specification of process etc.

There are many reasons for over processing, the most common reason is over designing a product leading to nightmare for the further processes to actually get to the required specification.

- **Defects**

Anything that client / Customer doesn't want

There's a famous saying that **" You always notice the lack of Quality"**
Any product which meets ones expectations is of good quality. In today's competitive environment there's not a single company which doesn't spend money on market research and analysis to capture the sentiments of customers and their needs. All these efforts are done to provide quality product as per the customer's expectations.

2nd Dimension of WCM: Work Environment (5-S):

Imagine an airport without any sign boards. One probably will need to arrive 6 hours before the departure if he/she is travelling for the first time from that airport. That's the power of visual management. The technique described ahead is one of the visual management tools implemented across various industries for better time management.

What is 5S?
This is a simple and systematic way of organizing and maintaining an efficient work area by eliminating waste.

5-S: Term refers to five Japanese words:

- Sort / Organize **(Seiri)**
- Set in Order / Neatness **(Seition)**
- Shine / Cleaning **(Seiso)**
- Standardize **(Seiketsu)**
- Sustain / Discipline **(Shitsuke)**

1st S - Sort:

- Separate needed from unneeded items.
- Remove unneeded items from the work area

Categorize the items as needed weekly / monthly / half yearly. Keep them in locations marked as per time period for visual management. Observe the consumption period of consumables, Quantity consumed daily and maintain stock levels as per that.

Before

After

In the above example the difference between "Before" and "After" picture is:

Before:
1) The Beams are placed perpendicular to the flow.
2) Different sizes are stacked over one another making it difficult to fetch the correct beam.
3) The beams are placed on wooden planks exposing the worker to unsafe work environment.

After:
1) The Beams are in the direction of work flow
2) The Beams of same size are stacked over one another.
3) The Beams are places on a well-designed platform to prevent falling of beam from stack.

2nd S – Set in order:

- Arrange needed items so they can be easily retrieved, used and put away.
- Reclaim unneeded space.

Before **After**

Templates for checking rolled cans needs to be sorted from box

Stand for holding templates so that sorting is avoided

In the above example, placement of the templates on vertical stand freed up the space occupied by the box. Secondly effort needed to identify the templates by taking out one by one from the box has been eliminated by placing them in visual identifiable position.

3rd S – Shine:

- Identify what needs to be cleaned.
- Identify when cleaning should be done.
- Assign responsibility for identified areas.

Before

After

In the above example, the store which was in mess has been cleaned and arranged properly with better lighting and labels on the rack to ease the retrieval of items.

4th S – Standardize:

- Prepare cleaning schedules and inspection check lists.
- Maintain records.

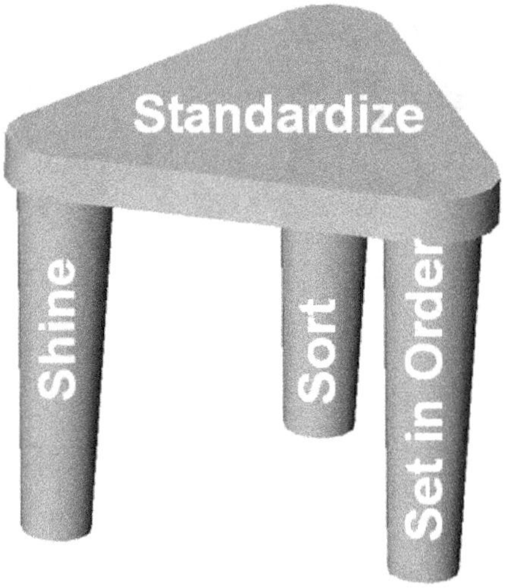

The above picture very much explains the crux of the first four steps of 5S. If any one of the legs will be missing then Standardization cannot be achieved. Standardization can be achieved by making visual SOP's/ SWI's (Standard Operating Procedure / Standard Work Instruction) for the tasks and displaying it near the work area. **Given below is the example of Visual SOP**:

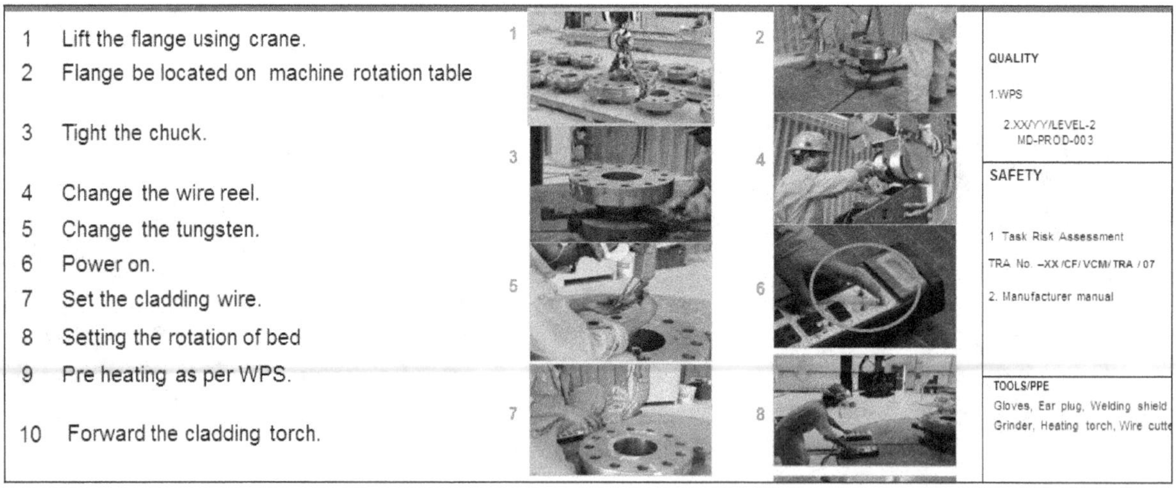
1 Lift the flange using crane.
2 Flange be located on machine rotation table
3 Tight the chuck.
4 Change the wire reel.
5 Change the tungsten.
6 Power on.
7 Set the cladding wire.
8 Setting the rotation of bed
9 Pre heating as per WPS.
10 Forward the cladding torch.

QUALITY
1.WPS
2.XX/YY/LEVEL-2
MD-PROD-003

SAFETY
1 Task Risk Assessment
TRA No. –XX /CF/ VCM/ TRA / 07
2. Manufacturer manual

TOOLS/PPE
Gloves, Ear plug, Welding shield
Grinder, Heating torch, Wire cutt

5th S – Sustain:

- Maintain the new standards of the work place organization.
- Perform periodical audits.
- Make 5S as organization culture.

Attached below is sample 5S audit checklist in order to sustain and maintain 5S culture in the organization.

Workplace 5S Audit Sheet

No	Category	Topic	Description & Score						Score /5	Improvement Points
			Audit questions:	1	2	3	4	5		
1	Sort	Workplace condition	Number of out of place items in the work area	15<	10 to 15	5 to 10	0 to 5	0		
		Removal of out of place items	A process for removing out of place items	Verbally assigned responsibilities	Schedule existing but no record of evidence like log book	Daily schedule existing and responsibilities assigned, log book is maintained	Visual display of the schedule in the area	Schedule Visual displays are up to date		
2	Set in order	Workplace condition	Number of necessary items not at point of	15<	10 to 15	5 to 10	0 to 5	0		
		Proper arrangement of items	A process for setting items in order	Items are not arranged and lie in the area randomly	Dedicated stands, floor space,shadow boards, file cabinets etc. for 50% items	Dedicated stands, assigned areas existing for putting tools, equipments, drawings, documents for 100% items	Visual display of the capacity, number of items, to be put on the stand / floor index of drawings displayed on the file storage location	Visual displays are up to date		
3	Shine	Workplace condition	% Area Free of trash, scraps, soil, leaks,	0%-20%	21%-40%	41%-60%	61%-80%	81%-100%		
		Housekeeping, Tool maintenance	A process for daily cleaning	Verbally assigned responsibilities	Schedule existing but no record of evidence like logbook, excel sheet	Daily schedule existing and responsibilities assigned, log book, excel sheet is maintained	Visual display of the schedule in the area	Schedule Visual displays are up to date		
4	Standardize	Workplace condition	% of processes for which standard	0%-20%	21%-40%	41%-60%	61%-80%	81%-100%		
		Revision of standards	A process to review, revise and update	No auditing,	Schedule developed for auditing of existing	SOP's existing and followed but not displayed at the point	Visual display of SOP's for all the processes in	Visual displays of SOP's are up		
5	Sustain	Workplace condition	System for maintaining current	No refrence of 5S in tool box	5S is included as part of daily tool box talks.	Teams showing good 5S scores are awarded and	Area Visit Ready	Area Visit Ready		
		Auditing 5S	A process for auditing and checking 5S	No auditing,	Schedule exists and followed, but no	Schedule exists and followed, record of evidence	Publishing audit scores and 5S projects on LAN	Establishing 5S scores as one		
			5S score							Auditor: Audit date: / /

5S is not only applicable for operations but more so for office environment as well. A 2005 study commissioned by Stamford, a maker of office labelling systems, of Connecticut (USA)identified that:

- Each lost document costs the company $120
- 51% of those surveyed identified a link between employees organizational skills and job performance i.e. tidiness and productivity.
- there is around 40% waste in office operations

Considering so much of saving potential it becomes mandatory for all organizations to implement 5S in all the areas of the company. The best way to implement 5S is to implement it in one of the area by making small focused group and then sharing the benefits with other areas in order to motivate them to deploy it. The deployment of external consultants or change agents is not much beneficial since the general feel among the employees is that it is the consultants responsibility to get it implemented which in fact beats the whole purpose of the program at the beginning itself that is to involve the employees and bring about a cultural change.

I.E.s are involved in improving the OEE (Overall Equipment Effectiveness) of the equipment which is the fourth pillar of the WCM.

4th Dimension of WCM: Equipment Effectiveness / TPM:

The need for improving OEE generally occurs when the workshop gets large orders and the objective is to squeeze every bit of money invested in the equipment sitting on the shop floor.

However the improvement of OEE or any improvement for that matter should be planned in the lean period to get the desired results.

OEE is the product of three variables i.e.

1) Availability Rate
2) Performance Rate
3) Quality Rate

Problem: A CNC machine is giving a finished product which has a cycle time of 60 seconds. In a shift of 8 hrs. it produced 310 finished products out of which 30 were found defective on inspection. The work force in the company is allowed to have Prayer time of 30 minutes, Lunch time of 30 minutes, TPM time of 10 minutes and meeting time of 10 minutes within a shift. That day during operation the machine broke down 4 times every B/D lasting 30 seconds, operator stopped the machine 7 times consuming total of 4 minutes, operator set up the machine 12 times each set up lasting 100 seconds. Find the OEE of the CNC machine.

Solution:

Let's start with calculation;

1) **Total Shift Minutes = 8 x 60 = 480 minutes**

2) **Planned /Allowable - Break/Rest Time = 80 minutes**
   - **Prayer Time = 30 minutes**
   - **Lunch Time = 30 minutes**
   - **TPM time = 10 minutes**
   - **Meeting Time = 10 minutes**

3) **Ideal Time for Production = 480 – 80 = 400 minutes**

4) **Time lost in operation that day:**
   - **B/D Time = 4 x 30 = 120 seconds = 2 minutes**
   - **Stoppage Time = 4 minutes**
   - **Set-up Time = 12 x 100 = 1200 seconds = 20 minutes**

   Unplanned Stoppage Time = 2 + 4 + 20 = 26 minutes

5) **Actual Production Time = 400 – 26 = 374 minutes**

Availability Rate = Actual Prod. Time / Ideal Production Time x 100
= 374 / 400 x 100 = 94%

6) **Expected Output based on Actual Production Time**
   = (374 x 60 seconds) / 60 secs.(cycle time) = 374 products

7) **Actual Output = 310 products**

Performance Rate = Actual Output / Expected Output x 100
= 310 / 374 x 100 = 83%

8) **Rejected Products = 30**

Quality Rate = (Actual Output – Rejected)/ Actual Output x 100
= (310 – 30) / 310 x 100 = 90%

Therefore OEE = Availability Rate x Performance Rate x Quality Rate
= 94% x 83% x 90% = 70%

Nowadays there are many software systems (e.g. Rhombus) which are hooked to the CNC circuit to capture the data automatically, the operator is provided with an LCD display in order to select the appropriate reason for any manual intervention to stop the machine. The OEE is then displayed on the machine for the shift.

The other important measurable related to the machines are MTBF (Meat Time Between Failure) higher the better and MTTR (Mean Time To Repair) lower the better. OTE (Overall Throughput Effectiveness)

4 APPLICATION OF I.E. IN AUTOMOBILE SECTOR

Automobile Sector is one of the most innovating sector. The Lean Manufacturing, Toyota Production System and Just in time concepts all originated in this sector and spread to other areas like Finance, Manufacturing, Construction and BPO's. The transition of technology in automobile sector is so swift that if the company is not agile enough to get up to speed then it can lose to competition and may finally wither away. Then there are vehicle regulatory norms which varies from country to country, therefore the automobile company has to match the performance & quality to the best of the requirements in order to be saleable in foreign land.

QCD Trilogy

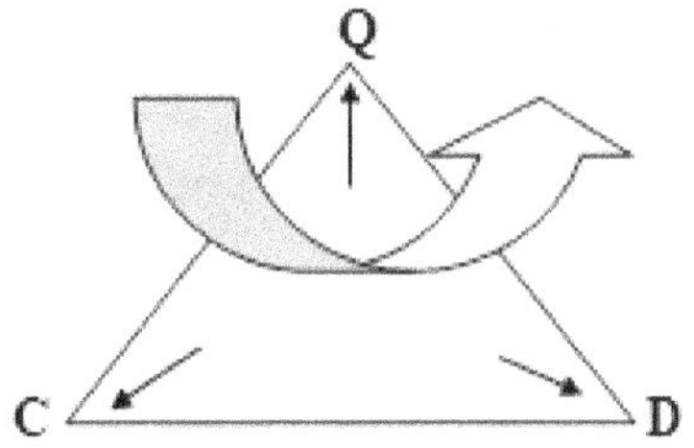

Initially the automobile sector had to keep a balance between **Quality, Cost and Delivery** but now a fourth factor which determines that a company will have edge over another is constant and consistent **Innovation** in the products.

QCDI Square

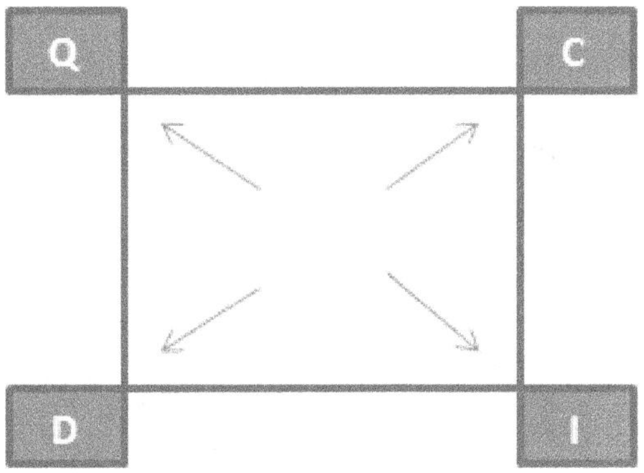

The role of an I.E. in automobile sector is limited to sustain and maintain the established systems, improve the existing processes, save cost through initiatives like Value Engineering, Manpower Management and Project management.

However it depends on person to person, in order to succeed in achieving cost cutting, improving processes the I.E. has all the freedom to take self-initiative. The I.E. must be up to date with the latest in the automobile technology and therefore must highlight to the management of the same.

The techniques of standard time calculation for a part assembly varies from one company to another. Some companies use PMTS (Predetermined Motion Time System) and others use MOST (Maynard Operation Sequence Technique). All these standards are inbuilt into the planning software systems (PLM) to design the shop floor layout and set the cycle time for various tasks.

Some of the areas that an I.E. is involved in Automobile Sector is:

1) Calculation of Standard Man-Hours for various components.

2) Value Engineering
    a. Scrap-less Blanking
    b. Part replacement with alternative material
    c. Removal of components without affecting core functionality of the vehicle.

3) Long term & short term manpower planning.
4) Competency mapping of the Permanent as well as Flexible workforce.
5) Overall and direct Productivity report formulation for MIS purpose.
6) Process improvement through initiatives in coordination with shop floor.

7) Implementation of Balance Score Card concept.

SMH (Standard Man-Hours):

Standard Man-hours of components form the basis of all current and future planning of an automobile company be it for Manpower planning, Strategic planning, facility Planning and Technology upgrade. All the major players in automobile sector have well established SMH for all components, the SMH's are reviewed periodically and modified if any improvement has taken place.

The Productivity of an Operation dept. is calculated based on SMH as:

> = (Total Components Produced, assembled x SMH for one component) / Target SMH based on the manpower provided or Equipment output in SMH for the shift

The Productivity report is one of the key output from I.E. dept. as part of the MIS (Management Information System)

One of the area where the SMH development is very complex is the automobile company's Central Equipment Manufacturing (CEM) shop where various types of Sheet metal Dies, Fixtures and Jigs are manufactured. In this area each product is different from another, there's no standardization as far as dimension and specifications are concerned. Therefore establishing SMH depends on large volume of time studies, along with capturing the specification & dimensions of the dies, fixture being manufactured to arrive at an empirical formula for SMH.

Given below is an example of such empirical equation to arrive at SMH of various types of fixtures:

Index:

Jigs/ Fixtures

Sr	Parameter required / Activity descr.	Code
1	O/L length of fixture/base plate	L
2	O/L width of fixture/base plate	W
3	Weight of fixture in kg	Wt
4	No of sub-assemblies	Nsa
5	No of fab-assys	Nfa
6	No of manufactured parts qty	Nmfd
7	No of standard parts qty	Nstd
8	No of dowels	Nd
9	No of cap-screws	Ncs
10	No of drill / slip bushes	Ndb

Sheet metal fixtures

Sr	Parameter required / Activity descr.	Code
1	Length of base plate/fixture	L
2	Width of base plate/fixture	W
3	Height of base plate/fixture	H
4	No of uprights	Nup
5	No.of gauge pins	Ngp
6	No.of L-blocks	Nl
7	No of nc blocks	Nnc
8	No of mfd parts	Nmf
9	No of cylinders	Ncyl
10	No of d.c. valves	Ndc
11	No of arms	Nar
12	No.of buttons	Nb

Inspection/component checking fixtures

1	Length of base plate/fixture	L
2	Width of base plate/fixture	W
3	Height of base plate/fixture	H
4	3d m/c'ing area	A3d
5	No.of plp points	Nplp
6	No.of a-class holes	Nah
7	No.of templates	Nt

Fixture groups

1 Drll/Ream/Tap/Chf/C'bore
2 2. Insp./Checking/Assy
3 M/c'ing -Turn/ Mill/ Grind/ Bore/ Hob/ Broach/ Lap/ Honning

Empirical Equation to calculate SMH for various activities involved in manufacturing of Jigs / Fixtures:

Jigs / Fixtures

Sr	Activity description		Work content estm. Norm (std man-hrs)
1.01	Gas cutting of plates		T = 0.12 x Nmfd + 2L
	Fabrication of fabr.assys/weldments		T = 0.85 x Nfa + 2L
1.02	Light m/c shop work	-mfd parts	T = 0.85 x Nmfd + sq.rt of Nmfd
		-std parts	T = 0.50 x Nstd + sq.rt of Nstd
1.03	Jig boring work on base plate (avg)		T = 0.15 x Nmfd + 2L
	Jig boring work for drill/slip bushes,other parts		T = 0.55 x Nb
1.04	Base plate -grind thick.on Nexus grinding m/c		T = LW / 0.16 + L
	-machine 2 reference faces		T = 2L
	-finish grind thickness		T = 4L
1.05	Fixture issue, study drg/tpl collect mfd/ std/ b/o parts, check/identification		T = 0.18 x (Nmfd + Nstd) + 2L
1.06	Marking/ layout, detail drg.study, calc.co-ord		T = 0.54 x(Nsa+Nfa) +4L +sq.rt (Nmfd+Nstd)
1.07	Deburring of mfd parts		T = 0.08 xNmfd + 0.2 x sq.rt of Nmfd
1.08	Drill/ c'bore/ transfer holes (drill/tap)/ drill/ream dowel holes on items & base plate for assy		T = 0.25 x Ncs + 0.16 x Ndh + 6L + 0.02 x Ncs
1.09	Assembly,checking, truing & levelling : dismantle, grind,flame harden, assemble parts as required		T = 0.80xfx(Nsa+Nfa)+sq.rt (Nsa+Nfa+Nmfd)+5L
1.10	Pess fit/position slip/drill bushes		T = 0.20 x Nb + 0.25 x (Nsa + Nfa) + 2L
1.11	Inspection- relative positions/dimnl accuracy and functional reqts		T = 0.56 x (Nsa + Nfa) x f + 2L f = 1.25 for required relative accuracy of +/- 0.01
1.12	Fixture tryout/ corrections & painting & despatch work		T = 0.15 x Total mfg time + 0.5 + 4L

Empirical Equation to calculate SMH for various activities involved in manufacturing of Component checking / Inspection Fixtures:

Component Checking/ Insp. Fixtures

Sr	Activity description	Work content estn. Norm (std man-hrs)		
1.01	Light m/c shop work for guide pins/bushes	T = 1.40 x Nac		
1.02	Base preparation	L <= 0.8	0.8 < L <1.2	L >= 1.2
	cutting and fabr. of sq.tubes, resting pads & lifters	3.12	4.21	5.37
1.03	Araldyte preparation, pouring, drying and finishing	2.36	3.26	4.16
1.04	Cutting/pasting ciba blocks w.r.t. compo.shape	T = (25.82 x A3d) x Cf + 2L		
1.05	Manual roughing to remove excess matl for Cnc	T = 0.2 x time for activity no.1.03 + 4L		
1.06	Cnc m/c profile & 3d surface on ciba blocks	T = 12.23 x A3d + 4.57 x Cf + L		
1.07	Finishing/polishing of ciba blocks	T = 0.25 x time for cnc m/c'ing + 4L		
1.08	Cmm checking for correct location and resting of compo. (gauge pins/bushes, plp points)	T = 1.25 + 0.2 x (Nac + Nplp) + 2L		
1.09	Fixing of plp points (drill/c'bore/sizing), gauge pins, bushes, clamps and templates	T = 0.82 x Nplp + 0.34 x Nac + Nt + 2L		
1.10	Painting and despatch work	T = 2.28 + 0.72 x Ap Ap = A3d+2H(L+W)		

Index for a Draw Die:

	Draw Dies	Blank Area M^2	0.15	0.25	0.50	1.00	1.50	2.00	3.20	4.00
Sr	Parameters required / activities	code	Small		Med		Large		S.Large	
1	O/L Length of die-set	L	0.80	1.10	1.40	1.80	2.20	2.40	3.05	3.25
2	O/L Width	W	0.70	0.80	1.00	1.40	1.60	1.80	2.20	2.40
3	O/L Height	H	0.40	0.60	0.75	0.85	1.00	1.15	1.25	1.25
4	Profile m/c'ing length of die	Lp	1.13	1.49	2.02	3.12	4.13	4.92	7.32	8.40
5	3D Area for m/c'ing of die	A3d	0.41	0.65	1.04	1.86	2.60	3.20	4.97	5.77
6	No of Shoulder Screws	Ns	2	2	4	4	6	6	8	10
7	No of Stub Pins	Nsp	4	6	8	12	16	24	32	40

Empirical Equation to calculate SMH for various activities involved in manufacturing of Draw Die:

Sr No	Activity description		Work Content Estimation Norm (SMH)		
1. Draw Dies		Length of die-set-->	L < 1	1 < L < 2	L > 2
1.01	Plane bottom surfaces of punch, die & b/h		T = 2.5+9.66xLW+2L	4+ 9.66xLW+2L	7.2+ 9.66xLW+4L
1.02	Cnc m/c profile on punch, b/h		T =3.88+2.104xLpx2 + L		
1.03	Cnc m/c 3d surface on punch, b/h & die		T = (40.14 x A3d + 15 x Cfcnc) x 2 + 2L		
1.04	M/c clamping pads, wear pads on P, D & b/h with anchoring slots		T = 10 + 2L	20 + 4L	44 + 8L
1.05	M/c top faces & cavities on punch & die		T = (1.8 x LW + 4L) x 1.5H		N=no.of parts
1.06	Die issue/ drg study/ collect parts/ check dimns		T = 2 +0.18N +2L	3.5 +0.22N +2L	5.5 +0.28N +4L
1.07	Die set making/layout/assy		T = 10 +2L	20 +2L	40 +4L
1.08	Stone & polish punch (master) & b/h		T = (5.25 +26.90A3d +7.5 xCfpol) xsqrt of L		
1.09	Drill air vent holes		T =1.67 x Flm	Flm = (6.52xA3d+3.46xCfcnc)	
1.10	Flame harden punch, b/h and die		T = 0.325xFlm	= flame harden length	
1.11	Polish punch, b/h & die after heat treatment		T = 0.2 x Stoning/polishing time x 2 +2L		
1.12	Stone/polish die before & after tryout		T = Time of 1.08 x 1.8		
1.13	Spotting die with punch and b/h		T = 0.4 x stoning/polishing time of punch & bh +2LW		
1.14	General asembly, painting & dispatch work.		T = 0.2x(Nsc+Nsp) +4LW +2.28+5x(L+W)		
1.15	avg. fabr. work	-mfd parts			
1.16	Thermo. patterns work	-patterns mkg			
1.17		-cnc m/c'ing			
1.18	Light m/c shop work	-std parts			
1.19		-mfd parts			
1.20	Jig boring work	-std parts			
1.21	Tryout-->Inspn-->Corrections-->Despatch		T = 0.4x cnc m/c'ing hrs +4LW		

Value Engineering:

Definition: The systematic application of techniques which identify the function of a product or service, establish a monetary value for the function, and provide the necessary function reliability and quality at the lowest overall cost. **(Society of American value engineers)**

Phases of V.E. :

1. **Orientation phase:**
   Identification of problems, selection of projects, formation of teams, laying down objectives and targets and in depth training of the teams.

2. **Information phase:**
   Collection of all relevant information like drawings, technical specs, manufacturing processes, detailed cost break-up, quality, procurement and production problems.

3. **Function analysis phase:**
   Analysis of function and classification into primary, secondary, essential and unnecessary functions, determination of function cost, worth, value gap & value index.

4. **Creative phase:**
   Application of brainstorming and other creativity techniques in order to generate a large no. of ideas for providing the functional requirement.

5. **Evaluation phase:**
   Critical evaluation, screening and short listing of the generated ideas using techniques like paired comparison method, evaluation matrix etc.

6. **Investigation phase:**
   In depth investigation and validation of short-listed ideas to arrive at a few optimum and practical solutions.

7. **Recommendation phase.**

8. **Implementation phase.**

Examples of V.E.:

1) Replacement of a sheet metal part with FRP (Fiber Reinforced Plastic) e.g. Now the fender (Bumper) of the cars which used to be metallic has been replaced with FRP in both back and front. The functionality remains the same as it has been designed in a way that it reduces the impact of accident to the main component of the car and the driver.

2) Decreasing the thickness of the metallic components to lighten the weight and thereby saving material cost.

3) Improving the process / component design to minimize scrap

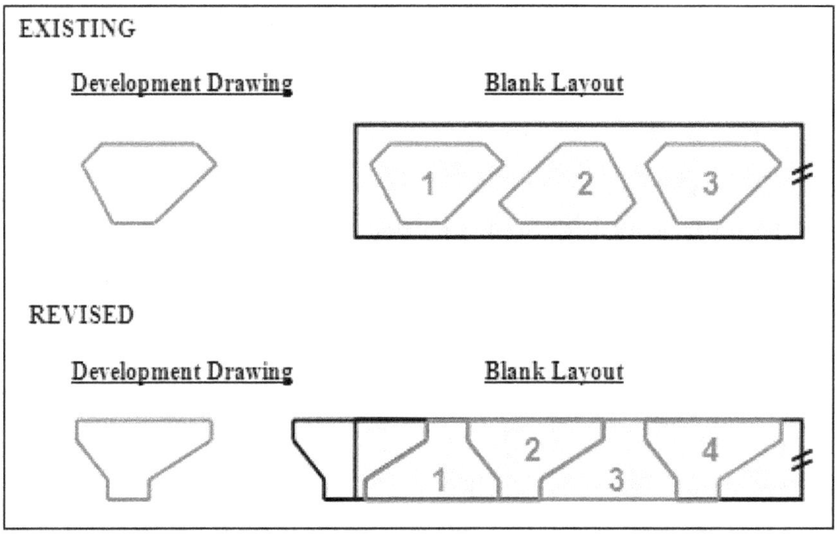

Balance Scorecard:

Definition: The Balanced Scorecard (BSC) is a strategic performance management framework that allows organizations to manage and measure the delivery of their strategy.

History:

The concept was initially introduced by Robert Kaplan and David Norton in a Harvard Business Review Article in 1992 and has since then been voted one of the most influential business ideas of the past 75 years.

In 1996, Kaplan and Norton published the book *The Balanced Scorecard*, and in the year 2000 they published the book *The Strategy Focused Organization*.

Perspectives:

The Four Perspectives:

- Financial Perspective
- Customer Perspective
- Internal Process Perspective
- Learning and Growth Perspective

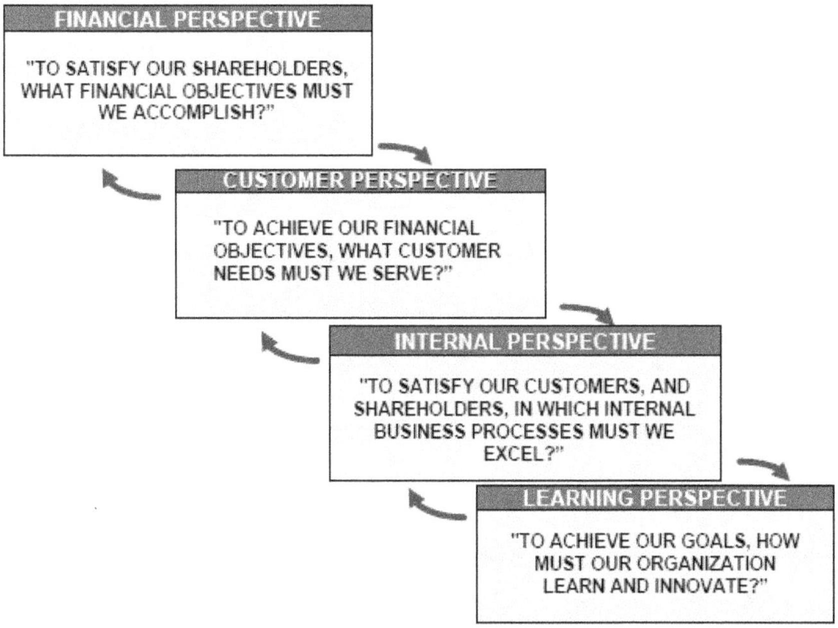

- **Financial performance** is a lag indicator – provides a definition for organisation's success. Strategy describes how it intends to create sustainable growth in shareholders' value.

- **Customer** – success with targeted customers provides a principal component for improved financial performance. Satisfaction, retention and growth are lag indicators, while customer value proposition is the central point of strategy.

- **Internal Processes** create / deliver the value proposition for customers. Performance here is a leading indicator of subsequent improvements in customer and financial outcomes

- **Learning &Growth** – intangible assets – ultimate source of sustainable value creation. Describe how people, technology & organisation climate are combined to support the strategy.

The objectives and targets set for an employee during yearly appraisal are linked to the strategic objectives that company has set for itself. This way people are aware as to how the tasks that they execute affect the company's strategy. The concept provides more of a transparency in the system if followed faithfully.

4 WORK MEASUREMENT TECHNIQUES

Why to Measure?

Measure of work brings knowledge. Through this knowledge, factual decisions & improvements can be made & control exercised.

Work Measurement:

Work Measurement is the application of the set of techniques intended to establish quantum of work to be done by an operator, in a given time, for a specified task under specified condition & at defined level of performance. Alternatively, it is scientific method of fixing standards of production

History of Work Measurement:

- Guessing
  The original form of Work Measurement.
- Historical Data
  Past Records were used to predicting future situation
- Stopwatch Time study
  'Frederic Taylor' emphasized on Work Element-Worker could be instructed as to best way to perform a task-Broken into element-Each element was studied to determine productive & useless part of element-Stopwatch study conducted for Productive element
- Motion Study
- Time & Motion Study (PMTS)

Time Study Process:

- Analyze the method.
- Break the operation into elements.
- Assess the time to perform each element.
- Performance rate the worker.
- Total the times to arrive at the normal time for the operation.

Performance Rating:

- Determination of pace of individual observed as compared to ideal, imaginary average worker working at 100% performance level of skill & effort
  - If the worker is performing at an above average level, the performance rating increases the time.

  - If the worker is performing at a below average level, the performance rating decreases the time.

Time Study: Advantages & Disadvantages

- Advantages
  - Fast to apply
  - Easy to learn
- Disadvantages
  - Subjective (performance rating)
  - Unable to use before production
  - Unable to simulate method changes

PMTS (Predetermined Motion Time System):

- A work measurement technique that utilizes catalogs (tables) of standard times which are assigned to fundamental motions that make up an activity.
- The times for the fundamental motions are totaled to determine the normal time for performing the activity.

History of PMTS:

- Pioneered by the micro motion studies of Frank and Lillian Gilbreth
  - Manual activities are combinations of basic elements.
  - Reduced motion = reduced time = higher productivity

MTM (Method Time Measurement):

- Developed by H. B. Maynard, G. J. Stegemerten, and J. L. Schwab in 1948
- Based on micro motions established by the Gilbreth.
- Identifies the variables which affect the time to perform each motion

MOST (Maynard Operation Sequence Technique):

- Considered as Revolutionary PMTS System
  - Introduced in early 1970's
  - Developed at H.B. Maynard & Co., Inc.
    - Kjell Zandin
  - Based on MTM data
  - Method Sensitive

Concept of MOST:

- Definition of Work
- Work is the displacement of a mass or object.
    - o Work = Force x Distance
- It's simply nothing but Movement of an Object
- There are only 2 ways for Moving the object
    - o Through Space
    - o Through restricted path
- It means **GET** and **PUT the Object**
- For example, you can **lift** a box and **place** it down three feet away.
- This is Activity based Technique
- All the manual activities & motions are breakdown into the standard sequence models of MOST

Sequence Models of MOST:

- The sequence models show the series of events or phases that occur when moving an object or using a tool.
- GET – ABG, PUT- ABP, RETURN – A
- Each letter in the sequence model represents a parameter that helps describe the action.
- There are three types of sequence models:
- General Move ABG ABP A
- Controlled Move ABG MXI A
- Tool Use ABG ABP _ ABP A

MOST system families:

Basic MOST – General Application
Mini MOST – Repetitive, Short cycle operation
Maxi MOST- Non- Repetitive, long cycle operation

MOST Analysis:

- Break down the task in method steps
- Body motions are organized into "Sequence Models" which describe

Activities(Method Steps)

- Each method step involves the movement of an object:
  - Freely Through Space GENERAL MOVE
  - Over a Restricted Path CONTROLLED MOVE
  - Using a Tool TOOL USE
  - Apply the appropriate index values to each parameter in the sequence model.
  - After putting index values to all the parameters do the calculation for time.

Time Study Vs MOST:

Time Study	**MOST**
• Low Accuracy Level • Disputes • Study output always doubtful • Time consuming • Applicators Experience plays crucial role • Speed dependent • Less faith level • Rating factor • Expensive Study • Need actual Demo	• Transparent • Easy to understand/absorb • Universal approach • More Accurate • No rating factor • Can analyze time without having any demos • Fast to apply • Less documentation • Method sensitive

5 MOST

MOST in Detail:

General Move:

Spatial displacement of an object. The object follows an unrestricted path through the air.

Examples of General Moves

- Pick up an object and place it on a shelf
- While holding a part, walk to a bin and drop part in
- Take a washer from your pocket and place it on a bolt

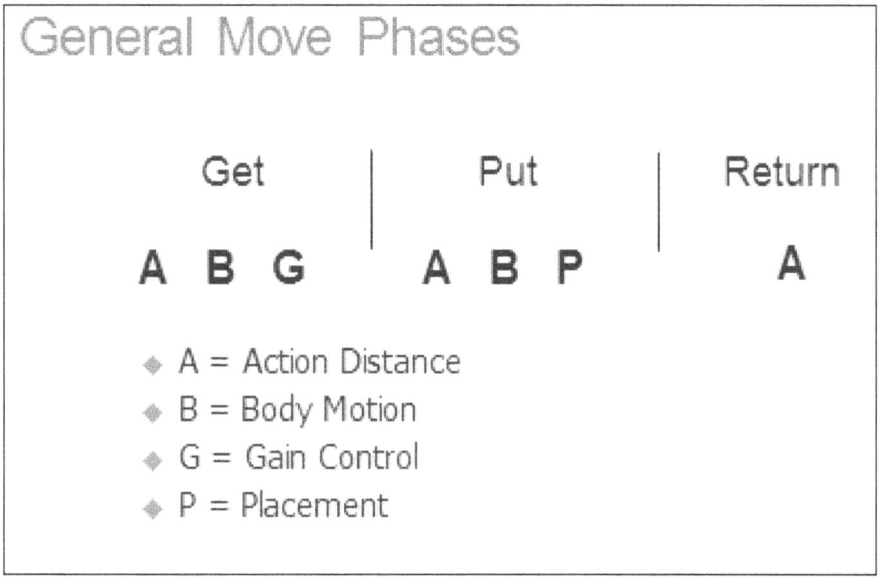

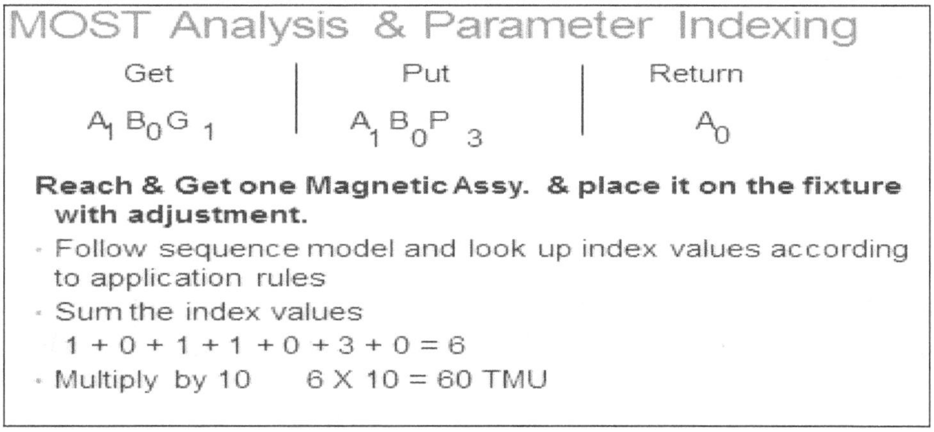

Time Measurement Units

1 TMU = 0.00001 hour
1 TMU = 0.0006 minutes
1 TMU = 0.036 seconds

1 hour = 100,000 TMUs
1 minute = 1,667 TMUs.
1 second = 27.8 TMUs.

Example:

A1 B0 G1 A1 B0 P3 A0 = 6 X 10 = 60 TMUs

This activity would take about 2.16 seconds.

Data Card:

ABG Get **ABP** Put **A** Return — **General Move**

Index x 10	A Action Distance	B Body Motion	G Gain Control	P Placement	Index x 10
0	≤ 2 in. (5 cm.)	No Body Motion	No Gain Control Hold	No Placement Hold Toss	0
1	Within Reach		Grasp Light Object Grasp Light Objects Simo	Lay Aside Loose Fit	1
3	1 - 2 Steps	Sit without adjustments Stand without adjustments Bend and Arise 50% occ.	Get Non-simo Get Heavy/Bulky Get Blind Get Obstructed Free Interlocked Disengage Collect	Loose Fit Blind Place with Adjustments Place with Light Pressure Place with Double Placement	3
6	3 - 4 Steps	Bend and Arise		Position with Care Position with Precision Position Blind Position Obstructed Position with Heavy Pressure Position with Intermediate Moves	6
10	5 - 7 Steps	Sit Stand			10
16	8 - 10 Steps	Bend and Sit Climb on Climb off Stand and Bend Through Door			16

A Action Distance Extended Values

Index	Steps	Distance (ft.)	Distance (m.)
24	11-15	38	12
32	16-20	50	15
42	21-26	65	20
54	27-33	83	25
67	34-40	100	30
81	41-49	123	38
96	50-57	143	44
113	58-67	168	51
131	68-78	195	59
152	79-90	225	69
173	91-102	255	78
196	103-115	288	88
220	116-128	320	98
245	129-142	355	108
270	143-158	395	120
300	159-174	435	133
330	175-191	478	146

Action Distance (A):

Action distance covers all spatial movements or actions of the fingers, hands and/or feet, either loaded or unloaded.

A_0 <= 2 inches

Any displacement of the fingers, hands, and/or feet a distance less than or equal to 2 inches. Example – placing nuts or washers on bolts located less than 2 inches apart.

A_1 Within reach
Actions are confined to an area described by arc of the outstretched arm pivoted about the shoulder
Example – All parts and tools can be reached without displacing the body by taking a step

A_3 one to two steps

A_6 Three or more steps

<u>Body Motion (B):</u>
Body motion refers to either vertical (up & down) motions of the body or the actions necessary to overcome an obstruction or impairment to the body movement.

B_6 Bend and Arise
From an erect standing position, the trunk of the body is lowered by bending from the waist and/or knees to allow the hands to reach below the knees and subsequently return to an upright position
Example – Pick up a box from floor and keep it on table

B_3 Bend and Arise, 50% occurrence
Bend & arise is required only 50% of the time during a repetitive activity, such as stacking or unstacking several objects

B_{10} Sit or Stand
When the act of sitting down or standing up requires a series of several hand, foot and body motions to move a chair or stool into a position that allows body to either sit or stand

B_{16} Stand and Bend / Bend and Sit
Occasionally a person sitting at a desk must stand up and walk to a location to gain control of an object placed below the knee level where bend and arise is required

B_{16} Climb On or Off

This parameter variant covers climbing on or off a work platform or any raised surface (approximately 3 feet or 1 meter high) using a series of hand and body motions to lift or lower the body

B_{16} Passing through door

Passing through a door normally consists of reaching for and turning the handle, opening the door, walking through the door and subsequently closing the door. This value will apply to virtually all hinge, double, or swinging doors. The three or four steps required to pass through the door way are included in the value.

Gain Control:

Gain control covers all manual motions employed to obtain complete manual control of an object and subsequently relinquish that control.

G_1 Light object

Example – Pick up a hammer from work bench, Obtain one washer from a parts bin full of washers

G_1 Light objects simo

Simo refers to manual actions performed simultaneously by different body members

Example – Pick up a bolt by left hand and washer by right hand simultaneously

G_3 Light object(s) Non SIMO

Because of the nature of the job or the conditions under which the job is performed, the operator is unable to gain control of 2 objects or of two suitable grasping point of one object simultaneously.

G_3 Heavy or Bulky

Control of heavy or bulky objects is achieved only after the muscles are tensed to a point at which the effects of the difficulty created by the weight,shape or size are overcome. We can identify this variant by the hesitation /pause needed for the attainment of sufficient muscular force require to move the object.

Example – Get hold on Battery located on the floor

G_3 Blind or obstructed
Example – Obtain a washer from a stud located on the other side of a panel (Blind)

G_3 Disengage
The application of muscular force is needed to free the object from its surrounding. Example – Remove the cap of marker. Tightly fitted socket from ratchet.

G_3 Interlocked
The object is intermingled or tangled with other objects & must be separated & worked free before complete control is achieved.
Example – Remove a hammer from crowded tool box

G_3 Collect
Gaining control of several object is accomplished
Example – Grasp a handful of washers from a bin

Placement (P):

Placement refers to actions occurring at the final stage of an object's displacement to align orient or engage the object with another before control of the object is relinquished. Placement includes limited amount of insertion (up to 2 inches) as part of the placement.

P_0 Hold
Placement does not occur

P_0 Toss object(s)
Example – Toss the waste paper into the dust bin

P_1 Lay aside
The object is simply placed in an approx location with no apparent aligning or adjusting motions
Example – Lay a hand tool aside after use

P_1 Loose fit
Example – Place a washer on a bolt

P3 Loose Fit Blind

The object is placed in more specific location than in Lay Aside, but the tolerances are so that very low mental, visual & muscular control will reqd.

P3 Place with Adjustments

Adjustments are defined as the corrective actions occurring at the point of placement caused by difficulty in handling object, closeness to fit, uncomfortable working condition, which recognize as a hesitation
Example – - Key in lock, Screw Driver on Screw head

P3 Place with Light Pressure

The application of muscular force is needed to seat the object,
for e.g. snapping action Example – Stamp on the envelope, Electric plug into the socket

P3 Place with Double Placement

Example – Place bolt through hole before placing nut

P6 Position with Care or Precision

Extreme care is needed to place an object, required high degree of Concentration, mental, muscular & visual co-ordination is reqd. Example – Thread a needle, Component soldering on PCB.

P6 Position Blind or Obstructed

Example – Place a nut on hidden bolt

P6 Position with Heavy Pressure

As a result of very tight tolerances, not weight of an object alone, a high degree of muscular force is needed to engage the object. Example – Position a book in book shelve.

P6 Position with Intermediate Moves

The additional handling is required to overcome awkward nature of the object
Example – Aligning the chair to a row by doing several sliding moves

Controlled Move

The movement of objects along a controlled or restricted path.

Controlled Move Activities

- Push a button
- Pull a lever 8 inches
- Flip a switch
- Press a clutch pedal

Parameters

Get	Move/Actuate	Return
A B G	M X I	A

- A =
- B =
- G =
- M = Move Controlled
- X = Process Time
- I = Alignment

All manually guided movements or actions of objects over a controlled path (Using fingers, hands, or feet)

Two categories:

- o Push/Pull/Pivot
  - push button, pull lever, open book, push cart
- o Crank
  - turn a hand wheel, turn the handle on manual pencil sharpener

Push/Pull/Pivot

M0 No Action

M1 Push/Pull/Pivot $\leq$ 12 inches (30 cm.)

Object displacement is achieved by a movement of the fingers, hands, or feet not exceeding 12 inches (30 cm).

Example – Engage feed on cutting m/c with short hand lever

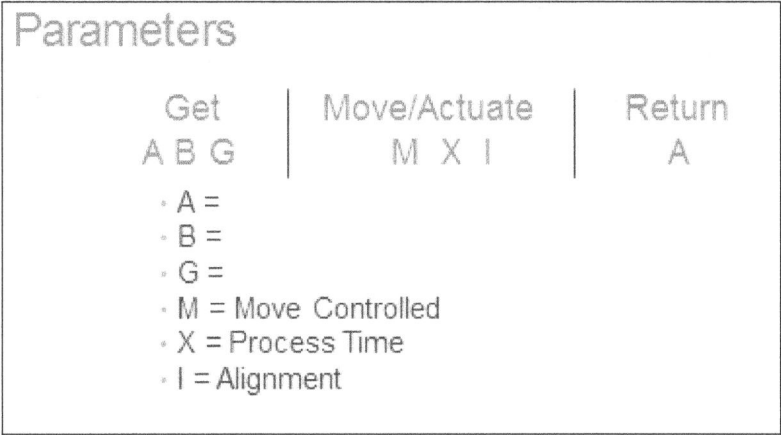

M1 Push Button

The device is actuated by a short pressing, moving, or rotating action of the fingers, hands, wrist, or feet (as long as no resistance is encountered).
Example – Press telephone hold button

M1 Push or Pull Switch

M1 Rotate Knob

M3 Push/Pull/Pivot > 12 inches (30cm.)

Object displacement is achieved by a movement of the hands, arms, or feet exceeding 12 inches (30 cm). The maximum displacement covered by this parameter occurs with the extension of the arm plus body assistance.
Example – Push the carton across conveyor roller

M3 Push/Pull with Resistance/ Seat/Unseat

This parameter variant covers the muscular force applied to "seat" or unseat" an object
Example – Engage the emergency brake on an automobile

M3 Push/Pull with High Control

Characterized by a higher degree of visual concentration, this parameter variant is sometimes recognized by noticeably slower movements to keep within tolerance requirements or to prevent injury or damage
Example - Turn the dial on a combination lock to a specific number
Push/Pull with 2 Stages $\leq$ 12 in. (30cm.) each
Push/Pull with 2 Stages $\leq$ 24 in. Total

Push/Pull with 2 Stages $\leq$ 12 in. (30cm.) each

An object is displaced in two directions or increments a distance not exceeding 12 inches (30 cm) per stage *without relinquishing control.*
Example – Shift 1st to 2nd gear with manual transmission

M6 Push/Pull with 2 Stages > 24 in.
An object is displaced in two directions or increments a distance exceeding 12 inches (30 cm) per stage without relinquishing control.
Example – Open & Subsequently close cabinet door

M6 Push with 1 - 2 Steps

M10 Push/Pull 3 - 4 Stages
Push/Pull with 3-5 Steps

M16 Push with 6 - 9 Steps

Crank
Cranking actions are performed by moving the fingers, hand, wrist, and/or forearm in a circular path more than half a revolution.

- M3 1 Revolution
- M6 3 Revolutions
- M10 6 Revolutions
- M16 11 Revolutions
- M24 16 Revolutions

Any motion less than half a revolution is not considered a crank and must be treated as a "Push/Pull/Pivot."

Push-Pull Cranking

- Back and forth movement of the elbow instead of pivoting at the wrist and/or elbow.
- This type of cranking is analyzed by using the number of pushes plus pulls as a frequency for the M1 parameter. (M3 if there is substantial resistance). Where possible, push-pull cranking should be replaced with pivotal cranking.

Process Time (X):

- The portion of work controlled by electronic or mechanical devices or machines, not by manual actions.
- Used for machine controlled activities having a fixed, short duration.

- X0 No Process Time
- X1 .5 Sec., .01 min., .0001 hr.
- X3 1.5 Sec., .02 min., .0004 hr.
- X6 2.5 sec., .04 min., .0007 hr.
- X10 4.5 sec., .07 min., .0012 hr.

  Note: Values are read up to and including.

Alignment (I):

- Manual actions following the Move Controlled or at the conclusion of process time to achieve an alignment or specific orientation of objects
- Eye Times are considered in the Alignment process.
  - Area of normal vision requires no additional eye time
    - Circle 4 inch (10cm) in diameter at 16 inches (40cm) distance
  - If one of the two points lies outside the area of normal vision, two separate alignments are required

Typical Objects

- I0 No Alignment
- I1 Align to 1 Point
- I3 Align to 2 Points < 4 in. (10cm.)
- I6 Align to 2 Points > 4 in. (10cm.)
- I16 Align with Precision (Several points)

Machining Tools

- I3 Align to Work piece
- I6 Align to Scale Mark
- I10 Align Indicator Dial

Non-typical Objects (Flat, Large Flimsy, Sharp, Difficult to Handle)

- I0 Against Stop(s)
- I3 1 Adjustment to Stop
- I6 2 Adjustments to Stop(s)
  1 Adjustment to 2 Stops
- I10 3 Adjustments to Stop(s)
  2-3 Adjustments to Line mark

Index the Sequence Model

Get	Move/Actuate	Return
A B G	M X I	A

- A press operator pushes two buttons simultaneously within reach to actuate the press. Press cycles for 2.5 seconds.

$A_1 B_0 G_1 \quad M_1 X_6 I_0 \quad A_0$

$= 1 + 0 + 1 + 1 + 6 + 0 + 0 = 9$

$= 9 \times 10 = 90$ TMU

Data Card:

ABG Get **MXI** Move/Actuate **A** Return — **Controlled Move**

Index x 10	M Move Controlled: Push/Pull/Pivot	M: Crank	X Process Time: Seconds	Minutes	Hours	I Alignment	Index x 10
0	No Action	No Action	No Process Time			No Alignment	0
1	Push/Pull/Pivot ≤ 12 in. (30 cm.) Push Button Push or Pull Switch Rotate Knob		.5 sec.	.01 min.	.0001hr.	Align to 1 Point	1
3	Push/Pull/Pivot > 12 in. (30 cm.) Push/Pull with Resistance Seat Unseat Push/Pull with High Control Push/Pull 2 Stages ≤ 12 in.(30 cm.) Push/Pull 2 Stages ≤ 24 in. Total	1 Rev.	1.5 sec.	.02 min.	.0004 hr.	Align to 2 Points ≤ 4 in. (10 cm.)	3
6	Push/Pull 2 Stages > 12 in. (30 cm.) Push/Pull 2 Stages > 24 in. Total Push with 1 - 2 Steps	2 - 3 Revs	2.5 sec.	.04 min.	.0007 hr.	Align to 2 Points > 4 in. (10 cm.)	6
10	Push/Pull 3 - 4 Stages Push with 3 - 5 Steps	4 - 6 Revs	4.5 sec.	.07 min.	.0012 hr.		10
16	Push with 6 - 9 Steps	7 - 11 Revs	7.0 sec.	.11 min.	.0019 hr.	Align with Precision	16

M Push or Pull Extended Values

Index	Steps
24	10-13
32	14-17
42	18-22
54	23-28
67	29-34

Crank Extended Values

Index	Revs.
24	12-16
32	17-21
42	22-28
54	29-36

X Process Time Extended Values

Index	Seconds	Minutes	Hours
24	9.5	.16	.0027
32	13.0	.21	.0036
42	17.0	.28	.0047
54	21.5	.36	.0060
67	26.0	.44	.0073
81	31.5	.52	.0088
96	37.0	.62	.0104
113	43.5	.72	.0121
131	50.5	.84	.0141
152	58.0	.97	.0162
173	66.0	1.10	.0184
196	74.5	1.24	.0207
220	83.5	1.39	.0232
245	92.5	1.54	.0257
270	102.0	1.70	.0284
300	113.0	1.88	.0314
330	124.0	2.06	.0344

Tool Use:

- Combination of General Moves and Controlled Moves
- Covers the Handling and Use of Tools
    - Get Tool
    - Put (Place) Tool
    - Use Tool
    - Put Tool Aside
    - Return Operator

Examples of Tool Use Activities

- Attach a nut on a bolt with 5 finger spins
- Fasten a screw with a screwdriver 8 wrist-turns
- Loosen a nut with a wrench using 3 wrist-strokes
- Cut open a box with a knife
- Record 4 digits on a paper
- Wipe a surface with a cloth

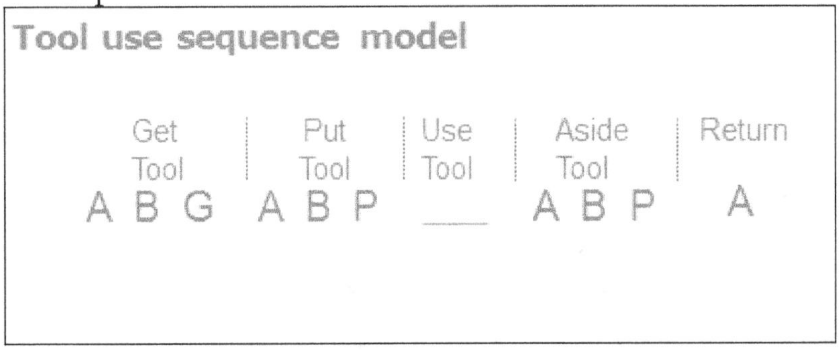

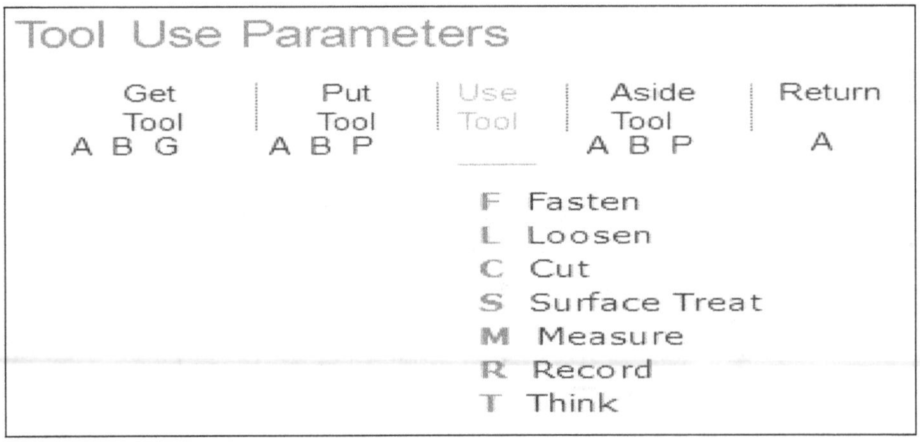

Tool Use Analysis

Get Tool	Put Tool	Use Tool	Aside Tool	Return
A B G	A B P		A B P	A

Pick up a screwdriver from the table within reach and place it on the head of a screw, turn down the screw 9 wrist-turns and set aside tool.

$A_1\ B_0\ G_1\quad A_1\ B_0\ P_3\quad F_{16}\quad A_1\ B_0\ P_1\quad A_0 \qquad = 240\ TMU$

Data Card:

Tool Use

ABG Get Tool · ABP Put Tool · * Use Tool · ABP Aside Tool · A Return

Index x 10	Fasten or Loosen (F, L)											Index x 10
	Finger Action	Wrist Action				Arm Action					Tool Action	
	Spins	Turns	Strokes	Cranks	Taps	Turns		Strokes	Cranks	Strikes	Screw Dia.	
	Fingers, Screw-driver	Hand, Screw-driver, Ratchet, T-Wrench	Wrench, Allen key	Wrench, Allen key, Ratchet	Hand, Hammer	Ratchet	T-Wrench 2-Hands	Wrench, Allen key	Wrench, Allen key, Ratchet	Hand, Hammer	Power Wrench	
1	1	-	-	-	1	-	-	-	-	-	-	1
3	2	1	1	1	3	1	-	1	-	1	1/4" (6mm)	3
6	3	3	2	3	6	2	1	-	1	3	1"	6
10	8	5	3	5	10	4	-	2	2	5		10
16	16	9	5	8	16	6	3	3	3	8		16
24	25	13	8	11	23	9	6	4	5	12		24
32	35	17	10	15	30	12	8	6	6	16		32
42	47	23	13	20	39	15	11	8	8	21		42
54	61	29	17	25	50	20	15	10	11	27		54

P Tool Placement

Tool	Index
Hammer	0 (1)
Fingers or Hand	1 (3 or 6)
Knife	1 (3)
Scissors	1 (3)
Pliers	1 (3)
Writing Instrument	1
Measuring Device	1
Surface Treating Device	1
Screwdriver	3
Ratchet	3
T-Wrench	3
Fixed End Wrench	3
Allen Wrench	3
Power Wrench	3
Adjustable Wrench	6

Tool Use Parameters:

Fasten (F) or Loosen (L)

Fasten or Loosen includes manually or mechanically assembling or disassembling one object to or from another using fingers, a hand, or a hand tool Index values for the F and L parameters are primarily grouped according to the body member performing the tool action.

- **Finger actions (Spins)**
  Include the movements of the fingers and thumb to run a threaded fastener down or out
  Example – Initial tightening of a nut with 3-4 finger spins

- **Wrist actions**
  Refers to the twisting motion of the wrist about the axis of the forearm OR The pivoting of the hand from the wrist with either a circular or back and forth motion
  Wrist Turn(s) Wrist Stroke
  Wrist Crank Tap

- **Wrist Turn(s) -** Tool actions using the hand, screwdriver, ratchet, or small T-Wrench
  Time for wrist turns includes
  -repositioning the hand after each action
  -final tightening or initial loosening of a fastener
  Example – tightening of a screw with a screwdriver with 3-4 wrist turns

- **Wrist Stroke -** Normally employed when using a wrench.
  Time for wrist stroke includes
  -wrench to be removed from & repositioned between the strokes
  -final tightening or initial loosening
  Example – using a wrench tighten a nut with 3-4 wrist strokes

- **Wrist Crank –**
  Tools that are rotated around a fastener while remaining affixed to it.
  Time for wrist crank do not include the time for final tightening or initial loosening of a fastener

- **Wrist Tap –**
  -The use of a small hammer, or other similar tools
  -short tapping motions performed with the hand as it is pivoted at the wrist
  Example – using a mallet apply 3-4 wrist taps

- **Arm actions**
  Include the motions of the hand requiring elbow and shoulder movements
  Arm Turn(s)
  Arm Stroke(s)
  Arm crank(s)
  Strike

- **Arm Turn(s)**
  Include use of a ratchet Arm turn time values includes time for final tightening or initial loosening.

- **Arm Stroke(s)**
  -applies to the normal method of using wrench
  -Arm stroke time values include final tightening or initial loosening activity

- **Arm Crank(s) –**
  -apply to tools used with a circular movement of forearm as it is pivoted at the elbow or shoulder
  -used with wrenches or ratchets
  -do not include the time for final tightening or initial loosening
  -The use of a hammer with arm action

Tool Use Example for Fasten/Loosen

1. Obtain a nut from a parts bin located within reach, place it on a bolt, and tighten with 7 finger actions

$$A_1B_0G_1 \mid A_1B_0P_3 \mid F_{10} \mid A_0B_0P_0 \mid A_0$$

$$(1 + 1 + 1 + 3 + 10) * 10 = 160 \text{ TMU} * 0.036 = 5.76 \text{ Sec}$$

2. Pick up a small screwdriver that lies within reach and fasten a screw with 6 finger actions and set aside the tool

$$A_1B_0G_1 \mid A_1B_0P_3 \mid F_{10} \mid A_1B_0P_1 \mid A_0$$

$$(1 + 1 + 1 + 3 + 10+1+1) * 10 = 180 \text{ TMU} * 0.036 = 6.48 \text{ Sec}$$

Tool Use Parameters - Cut, Surface Treat, Measure, Record and Think:

Cut(C)

-Manual actions employed to separate, divide, or remove part of an object using a sharp edged hand tool.

-covers the use of pliers, scissors, or knife for general cutting and related activities

Plier

Depending of the hardness of the wire material
and the diameter or gauge of the wire
Placement of the pliers is normally a $\mathbf{P_1}$

C_3 Soft

- Cutting a soft steel, copper, or other small gauge wire.
- Using the pliers with one hand and making one cut.

C_6 Medium

- Cutting a steel wire or cable
- Using the pliers with one hand and making two cuts

C10 Hard

- Cutting a heavier wire (approximately 10 gauge)
- Using the pliers with two hands and making two cuts

Scissors

- Cutting paper, fabric, light cardboard, or other similar material using scissors
- Index values are selected according to the number of cuts or scissors actions employed during cutting activity
  Examples – cutting of a thread with 1 cut cutting through a piece of fabric with four cutting actions

Placement of scissors is normally a $\mathbf{P_{1.}}$

Knife

- Cutting a corrugated box for opening

Example –
An operator picks up a knife from a workbench two steps away, come back, makes one cut across the top of a cardboard box, move two steps and keep aside the knife
$A_3B_0G_1$ $A_3B_0P_1$ C_3 $A_3B_0P_1$ A_0
(3+1+3+1+3+3+1) * 10 = 150 TMU * 0.036 = 5.4 Sec

Surface treat

- Covers only general cleaning activities performed with a cloth, an air hose, or a brush

Example –
Wipe metal sheet 4 sq ft using cloth from pocket
$A_1B_0G_1$ $A_1B_0P_1$ S_{32} $A_1B_0P_1$ A_0
(1+1+1+1+32+1+1)* 10 = 380 TMU * 0.036 = 13.68 Sec

Measure

- Includes the actions employed to determine a certain physical characteristics of an object by comparison with a standard measuring device
- Index values for M elements cover all actions necessary to place, align, adjust, and examine both the measuring device and the object during the measuring activity
- Placement will normally be analyzed with $\mathbf{P_1}$

$\mathbf{M_{10}}$ Profile gauge

- To compare the profile of the object to that of gauge
- The value includes placing and adjusting the gauge to the object, the visual

actions to compare the configuration of the object with that of gauge

M_{16} Fixed scale (12 inch scale)

- The value includes adjusting and readjusting the tool to two points and the time to read actual dimension from the graduated scale.

M_{16} Calipers <=12 inches

- Use of a Vernier caliper with a maximum measurement capacity of up to 12 inches
- The value includes setting the caliper legs to the object dimension, locking the legs in place and reading the Vernier scale

M_{24} Feeler gauge

- The use of a feeler gauge to measure the gap between two points
- The value includes fanning out the blades, reading and selecting the appropriate blade size, and positioning the blade to the gap to check for fit

M_{32} Steel tape <=6 feet

- This value is confined to the use of a steel tape from a fixed position and includes no walking between the two points to adjust the tape

M_{32}/M_{42}/M_{54} Micrometers <=4 inches

- Used for micrometers designed for maximum dimensions of 4 inches
- M32 for measuring depth, M42 for measuring outside diameter,
- M54 for measuring inside diameter
- Value includes setting the micrometer to the part, adjusting for fit, locking the device and reading the scale

M_{16} Plug gauge (go-no go)

Go + No go ends up to 1 inch

M_{24} Thread gauge (Plug or ring, go-no go)

internal or external threads up to 1 inch

M_{24} Vernier depth gauge

Up to 6 inches

M_{42} Thread gauge (Plug or ring, go-no go)
internal or external threads 1-2 inch

Record

- Covers the manual actions performed with a writing instrument or marking tool for the purpose of recording information
- Write
  to cover the routine clerical activities encountered in many shop operations
- Mark
  marking or identifying an object using a marking tool

The initial placement of a recording instrument before writing or marking usually occurs as a $\mathbf{P}_1$.

Example for record parameter

Write 6 digits on form using pencil from pocket

$A_1B_0G_1$ $A_1B_0P_1$ R_{10} $A_1B_0P_1$ A_0
(1+1+1+1+10+1+1)*10=160 TMU * 0.036 = 5.76 Sec

Get pencil from pocket, write date and signature on job card., keep pencil aside
$A_1B_0G_1$ A_1B_0 (P_1 A_0 R_{16}) $A_1B_0P_1$ A_0
(1+16)*2+1+1+1+1+1)*10=390 TMU * 0.036 = 14.04 Sec

Think

To cover only those types of reading and inspection activities that occur as a necessary part of a worker's job.

Inspect

Inspection work designed for making simple decisions Yes or no type defect inspection.

Read

Digits, letters, text of words Gauge reading

Normally the I.E. dept. develop excel templates for MOST and then for any activity they key in the index no. based on the type of activity. All of these when added provide the Man-Hours needed for assembly, manufacturing of the component.

6 APPLICATION OF I.E. IN OFFSHORE CONSTRUCTION

Application of I.E. concepts in Offshore construction industry is a big challenge in itself. I.E. dept. is in the nascent stages of establishing itself in the Offshore construction industry. The major responsibility of an I.E. is to convince the top management of the benefits that can be achieved if I.E. & lean concepts are deployed in the industry.

The very profound cliché that I.E. will get to hear in this sector is **" This is not manufacturing, these concepts will not work here".**

But this should not let I.E. down because when the management sees improved productivity, output & the workforce starts feeling the benefits of working effectively and productively on mostly value added task, I.E.'s starts getting the much needed support from the Top as well as from the people in operations.

Offshore construction is mostly project based sector where the company and the Project managers are bound by three factors i.e. Time , Cost and Quality. In most cases Cost factor is compromised to meet deadlines and provide quality product to the client on time because the clients' value stream is dependent upon the timely output from the contractor. Now to make up for this extra incurred cost, company goes for laying off employees, shutting down loss making /non-performing facilities and so on so forth.

This strategy works well in the short-term but the long term repercussions or the blind spots are hard to foresee. Some of them are:

- Lost reputation in the market
- Black listing by investors
- Lost faith in management by the clients and shareholders
- Lack of interest by people to work for such company leading to poor quality of human resource.

All this sends the company in forever spiral of instability, dearth of talent and hence low quality output.
These blind spots in fact more than ever before stresses the important role that I.E.'s can play in turning around the fortunes of the company.

With support from management, working in an offshore construction company for an I.E. is like a blank canvas. The question in front of I.E. is **"Where to begin?".** The role of an I.E. to begin with is to develop implementation and Roll-out strategy for the I.E. concepts.

Following is one of the strategy that an I.E. can follow to get the I.E. dept. up and running in an offshore construction sector:

Phase-1 (3-6 months): Implement I.E. concept in a small focused area

1) Select one small area which have a more controlled set up and resembles manufacturing environment.

2) Conduct an I.E. awareness training of the Manager, Supervisors and Leader-man of that area.

3) Conduct workshop with this group in order to highlight the problems being faced in the area and ask them to prioritize these based on certain metric like time taken to resolve the problem, Cost benefits, Ease of implementation, Improvement in quality, minimal capital investment etc.

4) Choose 2-3 highlighted problems based on priority and take agreement and go ahead from the Area management and supervision to make a working group for resolving it.

5) Apply various I.E. tools to find out the root cause of the problems and come up with various alternatives to resolve the problem.

6) Conduct weekly meeting with the area management to brief on the progress of the improvement project.

7) Implement the solution, modify the SOP's, train the workforce on new methodology.

8) Conduct periodic audits in the area to check whether workforce is facing any issues with the new approach and whether expected benefits are being realized.

9) Make a project report and a presentation based on the project.

Phase-2 (6-12 months):

Once I.E. achieves success in one area however small it may be, half the battle is won. Phase-2 of the implementation deals primarily in PR work and publicizing the benefits to other area managers. The stepwise approach in phase-2 involves:

1) Conduct meeting with all area managers and showcase the project executed in the focused area and share the benefits, however small they may be.

2) If possible develop an I.E. orientation presentation for all the managers to brief them on the concepts.

3) Propose to establish an I.E. committee with representatives from all areas and a fortnightly meeting to share and discuss the issues and projects in hand.

4) Roll out a common training plan for workforce from all areas on lean concepts like 5S, Waste Elimination, Kaizen as a long term strategy to take the organization towards lean culture.

Once an I.E. committee is established, the same approach as phase-1 has to be applied in rest of the areas to get improvement projects.

Conduct I.E. committee meetings periodically, use the platform to upgrade the managers of any new technology, processes being followed to make the work more productive.

In this phase I.E. will be working in close coordination with all the area managers and supervisors to achieve benefits in projects taken up. By the time Phase-2 is about to get over, I.E. can see the change in level of acceptance of I.E. concepts by different areas. Initially all these projects will be executed by the various area considering them as separate from their core job but as the process will mature, this will become part of their core activities without them being realizing that change is happening.

Phase-3 (1-3 years):
In this phase I.E. will still be engaged in some of the unfinished improvement projects taken up in phase-2 and with new projects being added to the list by the day. This phase is quiet critical in all respects because:

1) The expectations from management will be increasing, now they would like I.E. to get involved in major turn key projects to achieve higher savings.

2) The issues/problems will not be limited to small areas but to a more higher organization wide level.

3) The issues which have been considered as part of the operation historically will now be seen as bottleneck in the productive functioning of the organization.

To achieve success will be difficult considering the parameters and complexity of the problems, but this is what I.E.'s are being hired for.

I.E.'s now have to work as part of the decision making teams on the project and give their inputs in various processes being followed spanning from design review process to constructability reviews to procurement process to storage and logistics.

Success in this phase in any of the project is crucial for I.E. dept. to justify its existence, last long and bring about cultural change in the organization.

7 I.E. PROJECTS IN OFFSHORE CONTRUCTION

The projects/ Jobs in offshore construction sector for an I.E. range from:

1) Reduction in Inventory cost.
2) Reduction in maintenance cost of lift and shift gears.
3) Tracking and traceability of semi-finished/finished product, Equipment, Lifting gear in the yard.
4) Improvement in OEE of an equipment.
5) Facility layout design
6) Process improvement
7) Implementation of 5S
8) Reduction in time required for review of the design
9) Reduction in cycle time needed to procure material

Given ahead are real world examples of I.E. initiatives

5S initiatives example:

Before

After

Fabricated a stand to place the torch, cables and related tools. Reclaimed the space by removing the box.

Before

After

Previous Method - During dye penetrate testing, penetrate was spilling on the floor resulting potential danger of slip and environmental hazard.
Present Method - A U shape box is fabricated. It covers the end of pipe and directs the penetrate spill.
Improvement - Improved safety and hygiene of work area.

Kaizen Initiatives:
Example-1

Previous material movement

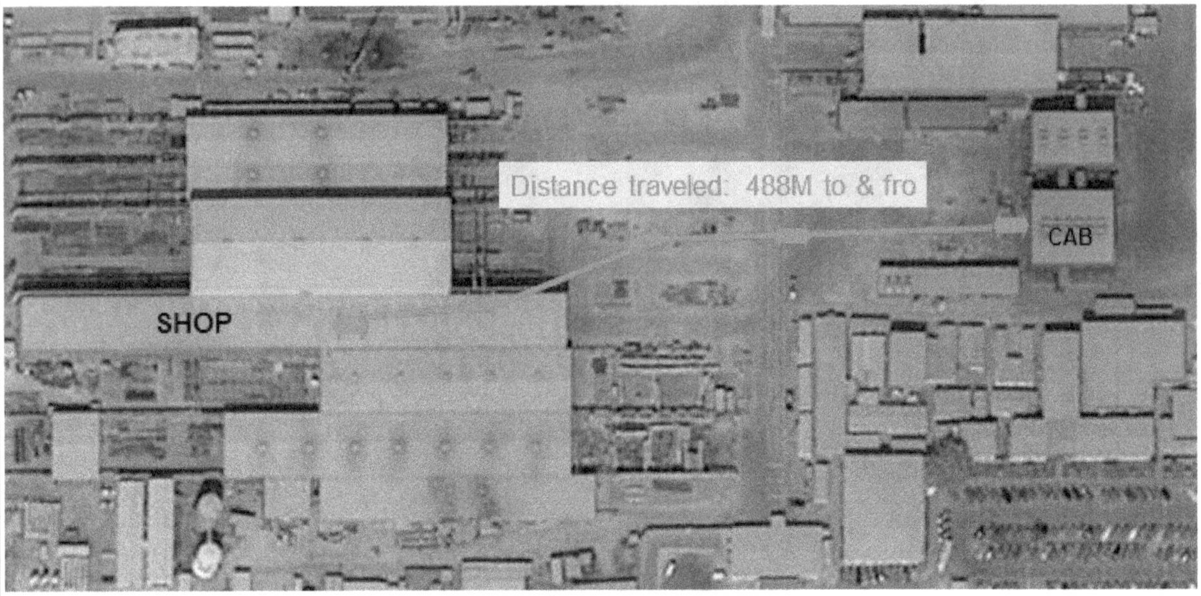

Present material movement

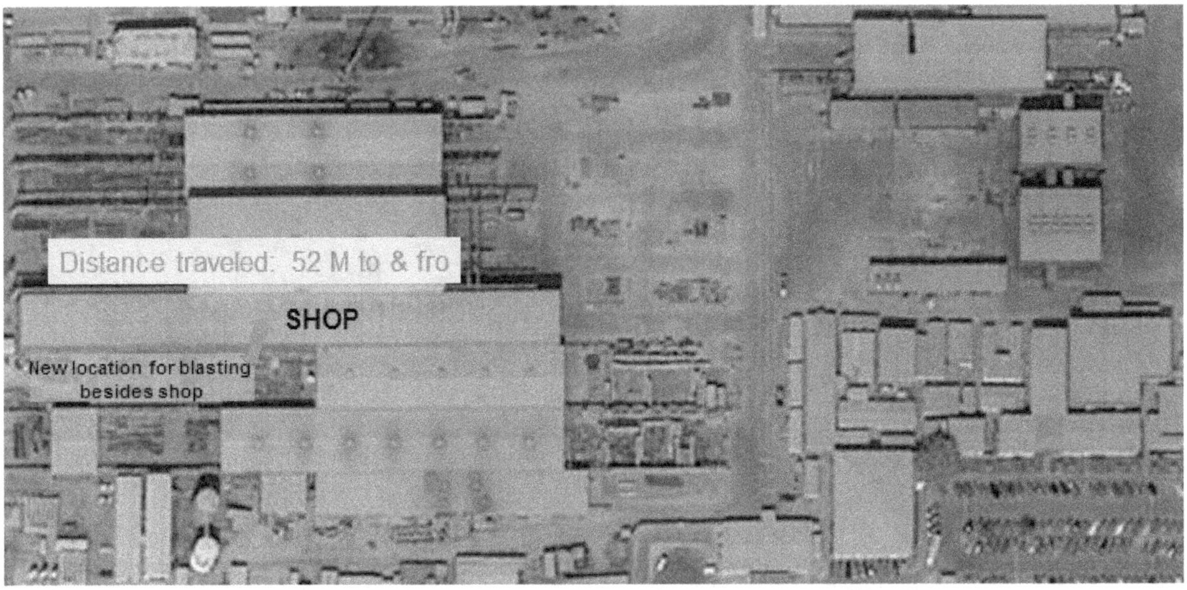

Example-2:

Before

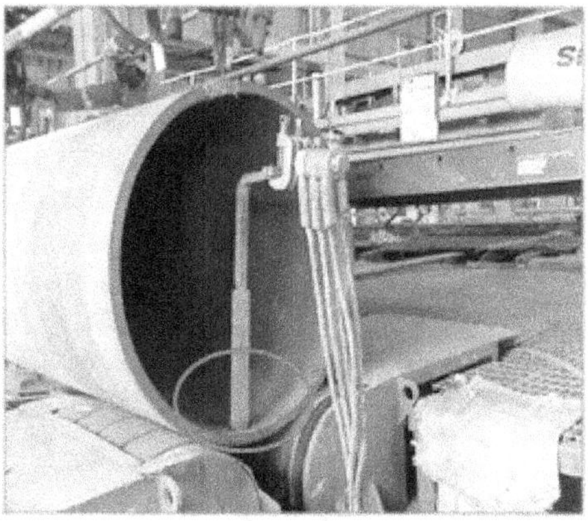

After

Previous Method - For **ground clamp** fixing, qualified stick welder was required to weld the clamp to the pipe. Once SAW is finished the clamp has to removed by gas cutting from the pipe. Finally grinding has to be carried out to smoothen the surface of pipe. Once grinding is completed MPI has to be carried out as per client specification.

Present method - In new set up no need for welding of ground clamp on the pipes. It is bolted system. The ground clamp is bolted and tighten with the pipe.

Improvements - Eliminates the need of welding, cutting, grinding and MPI of the clamp area on the internal surface of pipe. **Resulted in saving 3 man hours per length**

Example-3:

Before

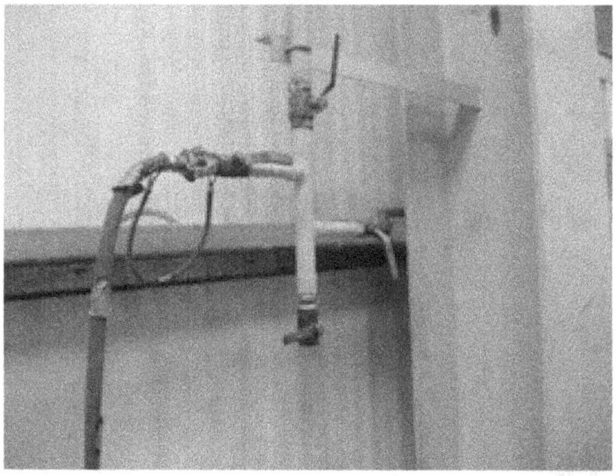

After

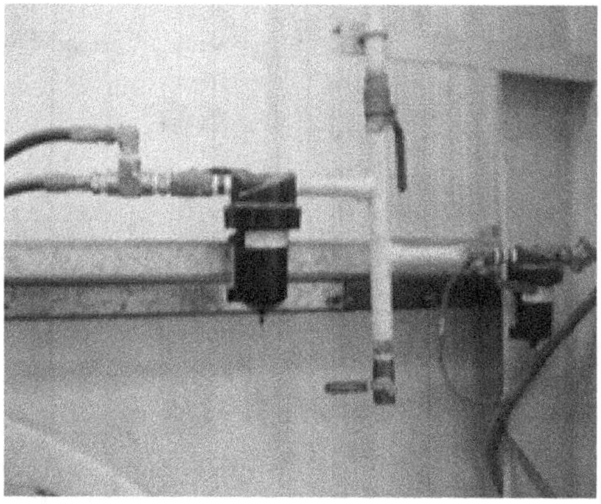

Concern:- CNC machines needs to be shut down for frequent maintenance.

Cause:- Moisture in the air utility lines.

Countermeasure:- Provide driers & filters in utility lines to prevent moisture going into CNC circuits.

Example-4:

Dual torch equipment is used to cut the beams into T-shape reducing set-up time and minimizing distortion.

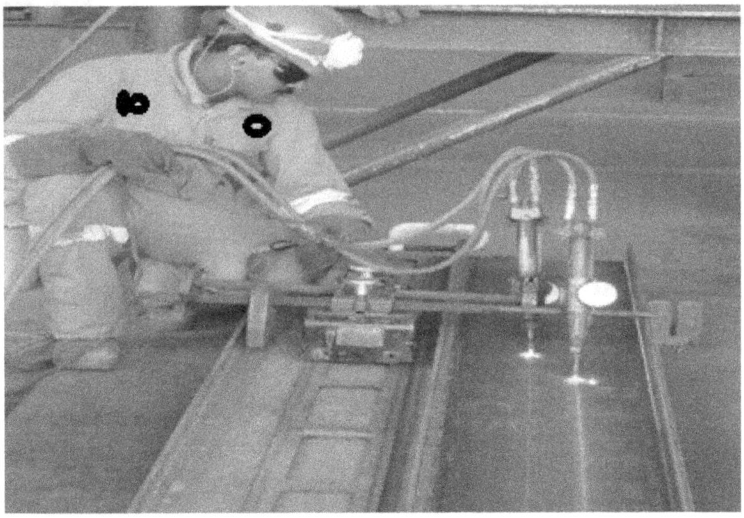

Savings: - For the total scope of XXX project i.e. 3000 meters of length to be cut. Projected savings will be (3000 X 3.6 X 2 = 21600 / 60 = 360 man hours)

Example-5:

Before	After

I.E. can suggest use of equipment for labor intensive jobs in order to reduce fatigue, simplify work and improve productivity of the workforce. I.E. should observe the activities and make a list of all such activities which require more manpower which otherwise could be done easily with use of mechanical aids.

Some of the labor intensive activities observed in Offshore construction industry are:

1) **Cable Pulling:** Pulling of cable around the corners in a congested environment on a deck / platform can be quite tiresome for workforce leading to days required for pulling certain length of cable.

The alternative way is to use electrically/hydraulically power cable drum and roller fitted in cable trays to lay the cable with less number of manpower.

a) Cable roller installed on the cable tray to ease cable pulling activity

b) Electrical Cable Drum rotating equipment

2) Shifting material on higher levels of the deck / platform:

Once the stacking of deck /platform is complete there is plethora of activities which needs to be done on a stacked deck like laying pipelines, welding structural supports, installing equipment, laying cables etc. to speak of few. The workforce is moving up and down carrying material, semi-finished and finished components to different deck levels. All these movements leads to fatiguing muscles, musculoskeletal injuries in the long run and delay in the job being executed. After stacking a deck stays (3-8 months depending upon deck complexity) in the yard for all these activity to finish before the deck sails away on the barge.

It makes all the sense to provide the structure with a hoist to the lift and transport men and material to different levels of the deck. It not only saves effort, man-hours but makes available more man-hours for productive jobs and a much needed relief for the workforce.

a) Lift installed on the platform

Tools to reduce work load of I.E.:

The task in construction field are long cycle and hence observing them whole day can make I.E. work full time on doing time studies only. Therefore camera can be mounted at the work station to record activities being performed. Later on the recorded video can be analyzed using software like **Timer Pro**.

There are 3D simulation software like **FlexSim** available in the market to design the facility and show simulation of the process being proposed. This tool provide the benefit of redesigning the facility depending upon the output expected by the management.

Value stream mapping software like **e-VSM and Visio** makes the mapping process all too easy and effective.

8 USEFUL CERTIFICATION & QUALIFICATION FOR I.E.'S

Some of the certifications and qualification that'll help add value to the knowledge pool of I.E.'s and may give that extra edge to stay ahead in the competition is one or a combination of the following:

1) Six Sigma Black Belt from ASQ (American Society of Quality)
2) Lead Auditor Quality Management Systems (Bureau Veritas - BVQi)
3) Project Management Professional -PMP(Project Management Institute)

There are distance learning & full time Master Programs available for I.E.'s to pursue, the area of interest varies from person to person. My personal preference is **"Masters in Operations Research"**, others may opt for **"Masters in Supply Chain & Logistics Management"** or **"Masters in Engineering Project Management"** or **"Masters in Manufacturing Management"**

Some on-hand experience on any good 3D simulation tool will prove to be handy. These courses are 2-3 days long classroom courses and can be learned within no time.

Industrial Engineers are expected to do anything and everything under the sun in the current employment scenario. Therefore I.E.'s should be open to anything and flexible enough to accept whatever is coming their way. The I.E. profession will give anybody an opportunity to work in any sector from Manufacturing to Automobile to Offshore construction to Aviation and much more because the ultimate goal of the profession is to measure, analyze and take decisions to help company achieve its goals by adopting best practices.

ABOUT THE AUTHOR

The author is qualified Industrial Engineer, Six Sigma Black Belt (ASQ) and Certified Lead Auditor Quality Management Systems (BVQi) with 11 years of work experience in wide variety of sectors like Manufacturing, Automobile, Offshore Construction and Aviation.

www.ingramcontent.com/pod-product-compliance
Lightning Source LLC
Chambersburg PA
CBHW081840170526
45167CB00007B/2853
9781495428104